Water Quality: Modelling, Monitoring and Treatment

Water Quality: Modelling, Monitoring and Treatment

Edited by Stephanie Fraser

SYRAWOOD
PUBLISHING HOUSE

New York

Published by Syrawood Publishing House,
750 Third Avenue, 9th Floor,
New York, NY 10017, USA
www.syrawoodpublishinghouse.com

Water Quality: Modelling, Monitoring and Treatment
Edited by Stephanie Fraser

International Standard Book Number: 978-1-68286-776-1 (Hardback)

This book contains information obtained from authentic and highly regarded sources. Copyright for all individual chapters remain with the respective authors as indicated. All chapters are published with permission under the Creative Commons Attribution License or equivalent. A wide variety of references are listed. Permission and sources are indicated; for detailed attributions, please refer to the permissions page and list of contributors. Reasonable efforts have been made to publish reliable data and information, but the authors, editors and publisher cannot assume any responsibility for the validity of all materials or the consequences of their use.

Trademark Notice: Registered trademark of products or corporate names are used only for explanation and identification without intent to infringe.

Cataloging-in-Publication Data

Water quality : modelling, monitoring and treatment / edited by Stephanie Fraser.
    p. cm.
Includes bibliographical references and index.
ISBN 978-1-68286-776-1
1. Water quality. 2. Water quality--Mathematical models. 3. Water quality--Measurement.
4. Water--Purification. 5. Water quality management. I. Fraser, Stephanie.
TD370 .W38 2019
363.739 4--dc23

TABLE OF CONTENTS

PREFACE

The measure of the biological, radiological, chemical and physical characteristics of water with respect to human consumption is known as water quality. Mining, urban runoff, construction and discharge of sewage are some common activities which pollute the water and affect its quality. Treatment of water focuses on eliminating impurities like organic chemical contaminants produced by industrial processes and petroleum use, microorganisms, inorganic contaminants like metals and salts, radioactive contaminants, and pesticides and herbicides. This book unfolds the innovative aspects of water quality modelling, monitoring and treatment which will be crucial for the progress of this field in the future. The aim of this book is to present researches that have transformed this discipline and aided its advancement. Those with an interest in this field would find this book helpful.

Various studies have approached the subject by analyzing it with a single perspective, but the present book provides diverse methodologies and techniques to address this field. This book contains theories and applications needed for understanding the subject from different perspectives. The aim is to keep the readers informed about the progresses in the field; therefore, the contributions were carefully examined to compile novel researches by specialists from across the globe.

Indeed, the job of the editor is the most crucial and challenging in compiling all chapters into a single book. In the end, I would extend my sincere thanks to the chapter authors for their profound work. I am also thankful for the support provided by my family and colleagues during the compilation of this book.

Editor

Review of the Impact on Water Quality and Treatment Options of Cyanide Used in Gold Ore Processing

Benias C. Nyamunda

Additional information is available at the end of the chapter

Abstract

Cyanide has been widely used in several industrial applications such as electroplating photography, metal processing, agriculture, food and the production of organic chemicals, plastics, paints and insecticides. The strong affinity of cyanide for metals such as gold and silver makes it suitable for selective leaching of these metals from ores. Cyanide is highly toxic; hence, there is a need to regulate and limit the amount of cyanide that may be discharged into the environment. Technologies focusing on the use of physical, chemical and biological methods have been developed to reduce the concentration of cyanide and cyanide compounds in wastewaters to permissible limits. This chapter reviews the current and emerging technologies for treatment of cyanide from wastewaters generated in gold mining processes.

Keywords: cyanide, gold, leaching, oxidation, wastewater

1. Introduction

Cyanide is an extremely toxic substance that is produced naturally and artificially. Cyanide has been widely used in several industries applications such as textile, plastics, paints, photography, electroplating, agriculture, metal treatment and mining. The high binding affinity of cyanide for metals such as gold, zinc, copper and silver has enabled it to selectively leach these metals from ores.

Southern Africa is a region rich in minerals such as gold. Countries in the southern Africa boost their economies through vast investments in gold mining. Cyanide leaching has become the dominant gold extraction technology since the 1970s replacing previously used less efficient and more toxic mercury. These gold mines discharge effluent containing toxic cyanides into natural water bodies posing the greatest threat to the quality of water intended for human use. Therefore, it is imperative to develop effective strategies for the removal of cyanide from aqueous industrial wastewater streams.

2. Gold extraction process using cyanide

Cyanidation is the predominant gold extraction technique since the late nineteenth century. The dissolution of gold in aqueous cyanide is commonly described using Elsner's equation [1]:

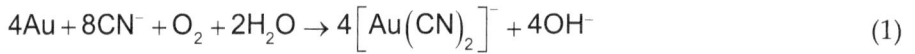

$$4Au + 8CN^- + O_2 + 2H_2O \rightarrow 4\left[Au(CN)_2\right]^- + 4OH^- \tag{1}$$

Gold dissolution is an electrochemical process in which oxygen is reduced at the cathodic zone, while gold is oxidised at anodic regions. The precise overall dissolution of gold in alkaline, aerated cyanide solutions taking place at cathodic and anodic regions is represented in Eqs. (2) and (3).

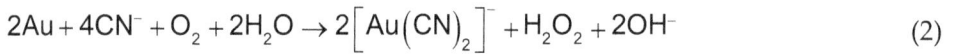

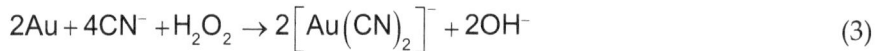

$$2Au + 4CN^- + O_2 + 2H_2O \rightarrow 2\left[Au(CN)_2\right]^- + H_2O_2 + 2OH^- \tag{2}$$

$$2Au + 4CN^- + H_2O_2 \rightarrow 2\left[Au(CN)_2\right]^- + 2OH^- \tag{3}$$

The main merits of cyanidation are the high selectivity of free cyanide for gold dissolution compared to other metals and an extremely high stability constant (2×10^{38}) of the gold cyanide complex [2].

Dilute sodium cyanide solutions within concentration ranges of 0.01–0.05% are used in mines for gold leaching [3]. Gold ore is subjected to physical processes such as milling, grinding and gravity separation prior to the addition of aqueous sodium cyanide to form slurry. The pH of the resulting extracting solution is increased by adding slaked lime or sodium hydroxide to prevent generation of toxic hydrogen cyanide [4]. The slurry pH is maintained at not less than 10.5 during cyanidation to prevent excessive loss of cyanide by hydrolysis through volatilisation of hydrogen cyanide. Oxygen an important component during cyanidation is continuously pumped into the slurry resulting in the formation of dicyanoaurate (I) complex.

Several methods are employed for cyanide leaching of gold ore [5]. However, agitation leaching is commonly used for most ores due to its commercial viability [6]. Leaching is typically done in steel vessels, and the solids are maintained in suspension by air or mechanical agitation.

The gold complex $NaAu(CN)_2$ is then extracted from leach solutions by adsorption onto solid adsorbents such as activated carbon or a synthetic ion exchange resin [7–11]. Activated carbon is the most commonly used adsorbent for gold extraction due to several favourable properties such as high adsorption capacity, good reactivation capabilities, low cost, readily available, high mechanical strength and wear resistance [12].

Gold complexes adsorbed onto activated carbon are eluted to produce concentrated high-grade gold solutions suitable for final gold recovery. Eluents such as sodium hydroxide [13] and organic solvents in aqueous solutions [14] have been used for desorption or stripping of gold from activated carbon.

Gold is extracted from solution into a concentrated solid form by a process termed recovery. Zinc precipitation [15] and electrowinning [16] have been used to treat concentrated gold solutions produced from activated carbon stripping. Eq. (4) represents the electrochemical reduction process for gold.

$$2\left[Au\left(CN\right)_2\right]^- + Zn + 4CN^- \rightarrow 2Au + 4CN^- + \left[Zn\left(CN\right)_4\right]^2 \tag{4}$$

The gold recovered from crude undergoes refining to produce crude bullion containing between 90 and 99.5% pure gold [17]. Refining involves roasting the crude gold to convert base metals such as iron, lead, copper and zinc to their respective oxides. This process is then followed by smelting, which removes base oxide impurities in form of slag. The bullion produced can be upgraded further to higher purity platinum group metals by processes such as pyrorefining, hydrorefining and electrorefining [18, 19]. These extraction processes leave behind toxic cyanide tailings.

3. Occurrence of cyanide in environment

Cyanide and related compounds are produced at low levels from plants such as sorghum, cassava, potato, broccoli, cashews and apricots [20]. Cyanide is found in certain bacteria, fungi and algae [21]. Anthropogenic sources of cyanide release also include smoke from cigarettes, automobile exhaust fumes and the production of acrylonitrile. Bulk occurrence of cyanide in the environment is attributed to the human operations in industries, metallurgical and mining activities. Cyanide is mainly produced industrially in form of hydrogen cyanide gas or solid sodium cyanide or potassium cyanide [22].

4. Forms of cyanide in aqueous solution

Compounds of cyanides present in water can be generally classified into total cyanide, complex cyanide and free cyanide [23–25]. These aqueous cyanide compounds exist as simple and complex cyanides, cyanates and nitriles. The most toxic form of cyanide is free cyanide, which exists either as cyanide anion or as hydrogen cyanide (HCN) depending on solution pH. HCN is predominant in aqueous systems at pH below 8.5 and can be readily volatilised [26, 27]. At higher pH values, the free cyanide is mainly in form of the cyanide anion. Aqueous cyanides form complexes with metal ions present in industrial wastewaters. These metallo-cyanide complexes exhibit different chemical and biological stabilities. The complexes are classified as weak acid dissociable (WAD) and strong acid dissociable [28, 29] in accordance with the metal-cyanide bond strength. Cadmium, copper, nickel and zinc form weak acid dissociable complexes that readily dissociate under acidic conditions [28]. Complexes of cyanide with cobalt, iron, silver and gold are strong acid dissociable (SAD). Both forms of complexes dissociate and release free cyanide. The stability of these complexes depends on several factors such as pH, light intensity, water temperature and total dissolved solids.

5. Toxicity of cyanide

Cyanide is extremely toxic to humans and aquatic life. Unlike toxic metal ions, the cyanide anion does not accumulate in the body, but instantaneously results in death of aquatic life and human beings in a short time at low dosages through depressing the central nervous system [30]. Cyanide strongly binds cytochromes inhibiting the electron transport chain in mitochondria and energy release in cells [31]. Liquid or gaseous hydrogen cyanide gains entry into the body through inhalation, ingestion or skin absorption. Exposing animals to hydrogen cyanide has several effects such as headaches, dizziness, numbness, tremor and loss of visual sharpness. Other toxic effects of cyanide include an enlarged thyroid gland, cardiovascular and respiratory problems.

6. Acceptable limits for the use of cyanide

There is need for the treatment of wastewater containing cyanide before discharging into the environment to protect water bodies. As a result of this, several countries and environmental bodies have imposed limiting standards for discharging wastewater containing cyanide to main natural water bodies. **Table 1** shows the set acceptable discharge limits of total cyanides by different organisations.

Agent	Cyanide limit	Reference
The U.S. Environmental Protection Agency (USEPA)	50 ppb (aquatic-biota)	[32]
	200 ppb (drinking)	
India Central Pollution Control Board (CPCB)	0.2 mg/L	[25]
Mexico	0.2 mg/L	[33]

Table 1. Permissible cyanide discharge limits in industrial effluents.

In view of the data outlined in **Table 1**, strategies aimed at cyanide recovery and removal need to be adopted to maintain concentrations within regulatory limits.

7. Cyanide removal strategies from wastewaters

Various cyanide attenuation processes have been successfully implemented in the treatment of industrial effluents. Gold mines in southern Africa have adopted various cyanide attenuation techniques aimed at reducing the toxin level in the tailings to internationally acceptable levels. The common methods of treating cyanide are natural, chemical and biological degradations [34].

7.1. Natural degradation of cyanide

Natural attenuation reactions occur in cyanide solutions placed in ponds or tailings resulting in the reduction in the cyanide concentration. The dominant natural degradation mechanism is volatilisation of hydrogen cyanide with subsequent atmospheric transformations to less toxic chemicals [34]. Other reactions such as hydrolysis, photolysis, oxidation, complex formation, oxidation to cyanate, thiocyanate formation and precipitation also take place. This natural process occurs with all cyanide solutions exposed to the atmosphere.

Cyanide forms complexes with metals ions in solution such as zinc, iron and copper. Ferri- and ferrocyanide complexes are extremely stable under most environmental conditions except when exposed to ultraviolet radiation [22]. Zinc and copper cyanide complexes are relatively unstable and can release free cyanide to the environment. Iron cyanide complexes are precipitated by several metals such as Zn, Cu, Ni, Pb, Sn, Cd and Ag over a wide range of pH (2–11). Cyanide and cyanide metal complexes adsorb onto clay, organic matter and oxides of aluminium, manganese and iron. The adsorbed cyanide can be naturally oxidised by oxygen, hydrogen peroxide and ozone into less toxic cyanate. The cyanate is hydrolysed under acidic conditions into ammonium salt and carbon dioxide. Cyanide can be biodegraded to ammonia, which is further oxidised to nitrate [35]. Under anaerobic conditions, HCN is hydrolysed to formic acid or ammonium formate as shown in Eq. (5).

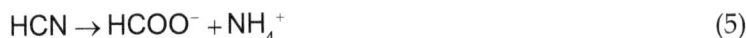

$$HCN \rightarrow HCOO^- + NH_4^+ \qquad (5)$$

Elemental sulphur and sulphur containing ores such as chalcopyrite react with cyanide to produce less toxic thiocyanate [36].

7.2. Chemical, physical and biological methods

Natural methods of cyanide attenuation have failed to produce effluents of acceptable quality. This has led to the development of numerous biological, physical and chemical treatment methods [37].

Among the methods used in removing cyanide from wastewater include photocatalysis [38], biotreatment [39], copper-catalysed hydrogen peroxide oxidation [40], ozonation [33], electrolytic decomposition, alkaline chlorination [22], reverse osmosis, thermal hydrolysis and adsorption [41]. Most of these methods have limited applications due to the high cost, production of toxic residues and incomplete degradation of all cyanide complexes [42, 43]. However, biodegradation of aqueous cyanide ions is cheaper than chemical and physical methods [30].

7.2.1. Chemical oxidation methods

7.2.1.1. Sulphur dioxide/air (INCO) process

This process was developed by a Canadian company, Inco metal limited, in 1984 [44]. The process makes use of air and sulphur dioxide in the catalytic oxidation of free and complexed cyanide to cyanate [37, 45, 46] as shown in Eq. (6). The process is catalysed by aqueous copper (II) ions under controlled pH of 8–10. The pH is normally maintained by addition of lime.

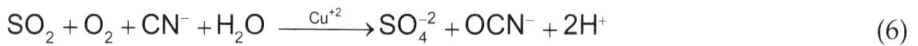

$$SO_2 + O_2 + CN^- + H_2O \xrightarrow{\ Cu^{+2}\ } SO_4^{-2} + OCN^- + 2H^+ \tag{6}$$

After completion of the oxidation process, previously metal ions complexed with cyanide such as Zn^{+2}, Cu^{+2} and Ni^{+2} are precipitated as metal hydroxides. This process effectively treats cyanide in slurries and solutions.

7.2.1.2. Alkaline chlorination

In this process, cyanide is oxidised by alkaline chlorine. The process converts all acid dissociable cyanide except for iron cyanide complexes and more stable metal-cyanide complexes. The process is a two-stage process. The first stage involves initial oxidation of free cyanide to cyanogens chloride followed by hydrolysis of cyanogens chloride to cyanate (Eqs. (7) and (8)) at pH 11.

$$CN^- + Cl_2 \rightarrow ClCN + Cl \tag{7}$$

$$ClCN + H_2O \rightarrow OCN^- + 2H^+ + Cl^- \tag{8}$$

During the second stage, cyanate is further oxidised to hydrogen carbonate and nitrogen as shown in Eq. (9). The reaction occurs at pH 8.5.

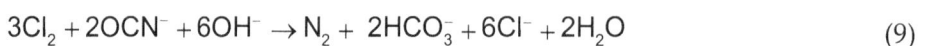

$$3Cl_2 + 2OCN^- + 6OH^- \rightarrow N_2 + 2HCO_3^- + 6Cl^- + 2H_2O \tag{9}$$

The alkaline chlorination process is primarily applied in the treatment of cyanide solutions rather than slurries, which consume a lot of chorine.

7.2.1.3. Hydrogen peroxide oxidation

Oxidation of cyanide tailings by hydrogen peroxide is more suitable for solutions rather than slurries. The oxidation process is maintained at pH of 9–10 to avoid formation of hydrogen cyanide [47]. The oxidation reaction is catalysed by copper (II) sulphate resulting in the production of carbonate and ammonium (Eqs. (10) and (11)).

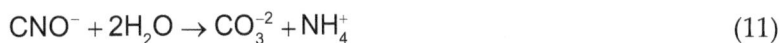

$$CN^- + H_2O_2 \xrightarrow{Cu^{+2}} CNO^- + H_2O \tag{10}$$

$$CNO^- + 2H_2O \rightarrow CO_3^{-2} + NH_4^+ \tag{11}$$

7.2.1.4. Ozonation

Ozone is a superior oxidant to oxygen and has been extensively studied in the oxidation of cyanide [48–51]. Two oxidation mechanisms of cyanide to cyanate by ozone have been proposed, namely simple (Eq. (12)) and catalytic (Eq. (13)).

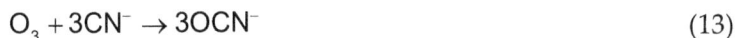

$$O_3 + CN^- \rightarrow OCN^- + O_2 \tag{12}$$

$$O_3 + 3CN^- \rightarrow 3OCN^- \tag{13}$$

Catalytic ozonation rarely occurs and has only been observed under high acidic conditions. Continued addition of ozone results in the formation of hydrogen carbonate and nitrogen (Eq. (14)).

$$2OCN^- + O_3 + H_2O \rightarrow 2N_2 + 2HCO_3^- \tag{14}$$

7.2.1.5. Peroxymonosulphuric acid

Peroxymonosulphuric acid (H_2SO_5) or Caro's acid [52] is used for cyanide treatment in gold tailings. Caro's acid is prepared in situ by the reaction of hydrogen peroxide with sulphuric acid since it easily decomposes. This acid is mostly used in situations where sulphur dioxide/air cannot be used. Caro's acid oxidises cyanide to cyanate as shown in Eq. (15).

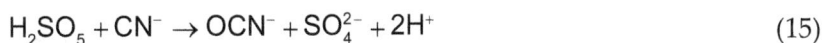

$$H_2SO_5 + CN^- \rightarrow OCN^- + SO_4^{2-} + 2H^+ \tag{15}$$

7.2.1.6. Precipitation of cyanide

Stable cyanide complexes can be precipitated by the addition of complexing agents such as iron. Iron cyanide complexes can coprecipitate other compounds containing cyanide in solution producing solids of cyanide salts. Finely divided insoluble iron sulphide is used for adsorbing free and complexed cyanide in solutions. The adsorption occurs at optimum pH of approximately 7.5. The iron sulphide is prepared from the reaction of iron (II) sulphate and sodium cyanide [53]. If hydrated ferrous sulphate is used, iron (II) cyanide precipitate is produced. Precipitation of iron cyanide occurs at pH between 5 and 6.

7.2.2. Physical methods

Cyanide tailings can be treated using physical methods such as dilution, membrane and electrowinning.

7.2.2.1. Dilution

This is a technique that does not destroy toxic cyanide, but dilute it with an eluent that reduces cyanide concentrations below acceptable discharge limits. Dilution is a cheap simple technique, which is often used as a standalone or in conjunction with other methods as a way of ensuring that discharged effluents are within permissible limits [54]. Dilution is normally an unacceptable method since it does not degrade or reduce the quantity of toxic cyanide exposed to the environment.

7.2.2.2. Membrane technology

Reverse osmosis and electrodialysis techniques using membranes have been used in extracting cyanide from wastewater. Both techniques have been effectively applied in the removal of free and complexed cyanide [55–57].

7.2.2.3. Electrowinning

Strong acid dissociable and weak acid dissociable cyanide complexes can be reduced to metals releasing free cyanide (Eq. (16)) by the application of an electric potential across electrodes immersed in complexes solution. The freed cyanide can then be treated by other processes.

$$\left[M\left(CN\right)_y \right]^{x-y} + xe^- \rightarrow M + yCN^- \qquad (16)$$

Four electrowinning cell designs have been developed for gold processing, namely Zadra, AARL, NIM graphite chip and MINTEK parallel plate cells [54, 58]. Electrowinning performs well in concentrated solutions and has been predominantly utilised for gold processing. This process is termed as Celec or HSA process [59] when it is used for cyanide regeneration.

7.2.2.4. Adsorption

Activated carbon, resins and minerals have been used for cyanide adsorption from solution. Contact vessels such as elutriation columns, agitation cells, packed-bed columns and loops have been used for this purpose. Various separation techniques such as floatation, gravity separation and screening are applied to remove the adsorbed cyanide from solution. The adsorbent is subsequently transferred into another vessel where cyanide is desorbed into low-volume solution, concentrated, reactivated and recycled.

7.2.2.5. Resins

Resins are typically polymeric beads containing a numerous surface functional groups capable of chelating or ion exchanging. Resins that require a substrate are deposited as thin film, while those that do work without a substrate are mostly used in continuous processes. The first column resin for cyanide recover was developed in 1959 [60]. Metal-cyanide complexes have been reported to adsorb more strongly to resins [61, 62]. The extent of adsorption depends on nature of resin used and how the resin and/or solution is pretreated [63]. Resins are cheaper and more effective than activated carbon since they resist organic fouling, have longer life, desorb faster and regenerate more efficiently [63]. Conventional, commercial strong base resins are most suitable for cyanide recovery since most common cyanide species in gold plant tailings are free cyanide anions within 100–500 mg/L range and the tricyano copper complex, both of which can be extracted directly from pulps using anion exchange resins [22].

7.2.2.6. Minerals

Free and metal-complexed cyanides are adsorbed by solid wastes, soils and ores containing minerals such as bauxite [$AlO.OH/Al(OH)_3$], ilmenite ($FeTiO_3$), haematite (Fe_2O_3) and pyrite (FeS_2). Mineral groups such as zeolites, clays and feldspars are also effective adsorbents [64, 65].

7.2.2.7. Activated carbon

Cyanide packed-bed systems can be used to adsorbed cyanide in dilute solutions [30, 66, 67]. Activated carbon has a relatively high affinity for many metal-cyanide complexes, including the soluble cyanide complexes of copper, iron, nickel and zinc [41, 68, 69]. Cyanide is adsorbed at various sites through chelation, ion exchange, solvation and coulombic interactions. This adsorption technique suffers a major drawback of being technically complex and expensive regeneration of activated carbon [70].

7.2.3. Biological oxidation methods

The use of microorganisms in the degradation of cyanide in tailing ponds has often been found to be potentially inexpensive and environmentally friendly compared to conventional chemical and physical processes [23, 71, 72]. Enzymatic activities associated with certain species of bacteria, fungi and algae are known to oxidise cyanide to less toxic cyanate [20, 73, 74]. Aerobic and anaerobic passive biological treatment processes are cost-effective alternatives to conventional cyanide treatment strategies since they do not need external energy, chemicals and routine maintenance. However, they suffer limitations such as the need for warmer climates (>10°C), large space and long retention times. **Figure 1** shows a flowchart for the aerobic and anaerobic oxidation of cyanides and thiocyanides in gold tailing ponds. Common passive biological treatment processes comprise engineered wetlands containing substrate or a mixture of organic and inorganic compounds like manure, straw, saw dust and limestone [22]. In anaerobic wetlands, bacterial oxidation of cyanides and thiocyanides to sulphates, carbonates and ammonia occurs as illustrated in Eqs. (17) and (18).

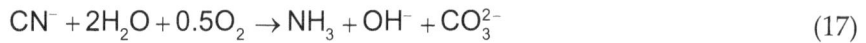

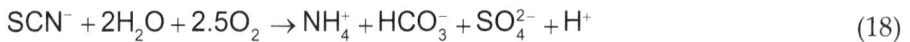

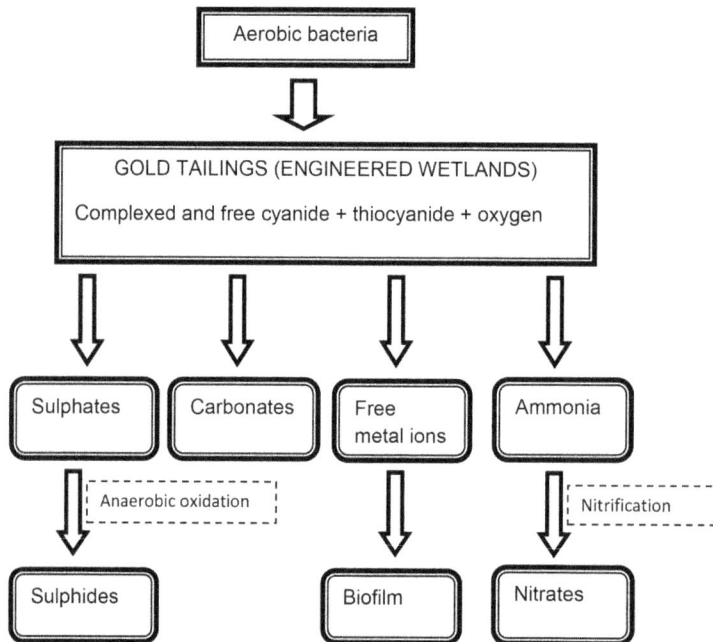

$$CN^- + 2H_2O + 0.5O_2 \rightarrow NH_3 + OH^- + CO_3^{2-} \tag{17}$$

$$SCN^- + 2H_2O + 2.5O_2 \rightarrow NH_4^+ + HCO_3^- + SO_4^{2-} + H^+ \tag{18}$$

Figure 1. Biological cyanide degradation processes in gold tailings.

The ammonia produced by the aerobic processes provides nutrients for microbial growth and the resultant uptake, sorption, conversion and/or precipitation of cyanides, thiocyanates, sulphates and nitrates by microorganisms [74]. The metals released during the oxidation of cyanide metal complexes are removed from gold tailings by chemical precipitation and/or adsorption on bacterial biofilm. Ammonia also undergoes further oxidation in a two-step nitrification process (Eqs. (19) and (20)).

$$NH_4^+ + 1.5O_2 \rightarrow NO_2^- + 2H^+ + H_2O \tag{19}$$

$$NO_2^- + 0.5O_2 \rightarrow NO_3^- \qquad (20)$$

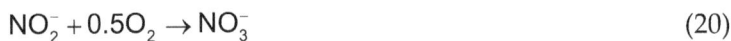

Sulphates undergo anaerobic reduction to sulphides (Eq. (21)). This process is effected by sulphate-reducing bacteria [75]. The sulphide produced is precipitated by metal ions resulting in its removal from aqueous tailings.

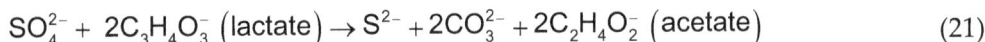

$$SO_4^{2-} + 2C_3H_4O_3 \, \text{(lactate)} \rightarrow S^{2-} + 2CO_3^{2-} + 2C_2H_4O_2^- \, \text{(acetate)} \qquad (21)$$

Several factors influence the biodegradation of cyanide in gold tailings. The most important environmental factors influencing biological treatment include pH, temperature, oxygen levels and nutrient availability. Enzymes that degrade cyanide are generally produced by mesophilic microorganisms, often isolated from soil, with optimum operating temperatures of between 20 and 40°C [34, 43, 76–79]. The availability of nutrient carbon has been found as a limiting factor in the biodegradation of metal-cyanide complexes [75].

Highly acidic and basic conditions have adverse effects on cyanide-degrading microorganisms since bacterial and fungal growth is optimal at pH 6–8 and 4–5, respectively [80]. Cyanide-degrading enzymes have optimum operating pH between 6 and 9. Concentrations of cyanide ions in water or slurries have an impact on the survival and growth of microorganisms. For instance, high cyanide concentrations have been reported to be toxic to *Klebsiella oxytoca* by damaging the nitrile-degrading enzyme, nitrile hydratase [81].

7.3. Emerging technologies on cyanide remediation

Since the 1990s, research has focused on introducing cyanide treatment technologies aimed at reducing costs and producing environmentally friendly products.

Carbon dioxide has been successfully used without a catalyst to replace SO_2 as an inexpensive alternative to the SO_2/air process [82].

Wastewater containing free and complexed cyanides can be oxidised by ultraviolet radiation in the presence of a semiconductor catalyst such as titanium dioxide [33]. When the catalyst mixed in the wastewater is exposed to the sunlight, it generates a highly reactive hydroxyl radical oxidant. These radicals initially convert cyanide to cyanate. Photocatalysis partially dissociates ferricyanide and ferrocyanide complexes to free cyanide and iron hydroxide. Photocatalytic oxidation is effective in relatively clear solutions. In the presence of ozone, ultraviolet oxidation does not produce undesirable by-products such as ammonia.

Solid or liquid cyanide wastes can be thermally decomposed upon treatment at elevated temperatures and pressure in batch or continuous mode [83]. This process is capable of destroying all cyanide complexes. Cyanide hydrolysis occurs in two steps (Eqs. (22) and (23)) producing ammonia and carbonates.

$$NaCN + NaOCl \rightarrow NaCNO + NaCl \tag{22}$$

$$2NaCNO + 3NaOCl \rightarrow H_2O + 3NaCl + N_2 + 2NaHCO_3 \tag{23}$$

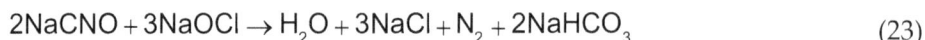

This cost-effective process was developed in the early 1990s for the treatment of wastes containing high concentrations of cyanide (100,000 mg/L). Thermal reduction reduces cyanide concentration to approximately 25 mg/L, which can be further oxidised by conventional methods such as ozone or hydrogen peroxide to environmentally permissible levels.

Free cyanide and cyanide complexes containing waste can be treated by electrochemical oxidation. This is an economical and environmentally friendly technique of destroying cyanide. The process results in cyanide ions being destroyed at the anode as metals are deposited at the cathode [27, 84]. During electrolysis, cyanide is initially oxidised at the anode-producing cyanate ions, which are further decomposed to carbon dioxide and nitrogen gas at the cathode (Eqs. (24) and (25)).

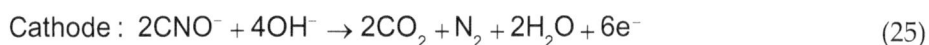

$$\text{Anode}: \ CN^- + 2OH^- \rightarrow OCN^- + H_2O + 2e \tag{24}$$

$$\text{Cathode}: \ 2CNO^- + 4OH^- \rightarrow 2CO_2 + N_2 + 2H_2O + 6e^- \tag{25}$$

8. Conclusion

There is a growing use of water as the gold mining activities increase. Water losses should be minimised and recycling adopted as much as possible. Cost-effective and environmentally friendly practices for cyanide treatment need to be implemented. The gold mining industry needs to implement the best practices for cyanide management that are aimed at assisting in protecting human health and reducing environmental impacts through discharge of permissible levels of cyanide in effluents into main water bodies. Such practice will ensure maintenance of good quality of water and sustenance of aquatic life.

Author details

Benias C. Nyamunda

Address all correspondence to: nyamundab@gmail.com

Midlands State University, Manicaland College of Applied Sciences, Department of Chemical and Processing Engineering, Mutare, Zimbabwe

References

[1] Thompson PF. The dissolution of gold in cyanide solutions. Transactions of the Electrochemical Society. 1947; 91(1): 41–71. doi:10.1149/1.3071767

[2] Donato DB, Nichols O, Possingham H, Moore M, Ricci PF, Noller BN. A critical review of the effects of gold cyanide-bearing tailings solutions on wildlife. Environment International. 2007; 33(7): 974–984. doi:10.1016/j.envint.2007.04.007

[3] Rees KL, Van Deventer JSJ. The role of metal-cyanide species in leaching gold from a copper concentrate. Minerals Engineering. 1999; 12(8): 877–892. doi:10.1016/S0892-6875(99)00075-8

[4] Jeffrey MI, Breuer PL. The cyanide leaching of gold in solutions containing sulfide. Minerals Engineering. 2000; 13(10): 1097–1106. doi:10.1016/S0892-6875(00)00093-5

[5] Miller JD, Wan RY, Mooiman MB, Sibrell PL. Selective solvation extraction of gold from alkaline cyanide solution by alkyl phosphorus esters. Separation Science and Technology. 1987; 22(2–3): 487–502. doi:10.1080/01496398708068965

[6] Örgül S, Atalay Ü. Reaction chemistry of gold leaching in thiourea solution for a Turkish gold ore. Hydrometallurgy. 2000; 67(1): 71–77. doi:10.1016/S0304-386X(02)00136-6

[7] McDougall GJ, Hancock RD, Nicol MJ, Wellington OL. Copperthwaite RG. The mechanism of the adsorption of gold cyanide on activated carbon. Journal of South African Institute of Mining and Metallurgy. 1980; 80(9): 344–356.

[8] Yalcin M, Arol AI. Gold cyanide adsorption characteristics of activated carbon of non-coconut shell origin. Hydrometallurgy. 2002; 63(2): 201–206. doi:10.1016/S0304-386X(01)00203-1

[9] Cortina JL, Warshawsky A, Kahana N, Kampel V, Sampaio CH, Kautzmann RM. Kinetics of goldcyanide extraction using ion-exchange resins containing piperazine functionality. Reactive and Functional Polymers. 2003; 54(1): 25–35. doi:10.1016/S1381-5148(02)00180-3

[10] Bachiller D, Torre M, Rendueles M, Díaz M. Cyanide recovery by ion exchange from gold ore waste effluents containing copper. Minerals Engineering. 2004; 17(6): 767–774. doi:10.1016/j.mineng.2004.01.001

[11] Fleming CA, Mezei A, Bourricaudy E, Canizares M, Ashbury M. Factors influencing the rate of gold cyanide leaching and adsorption on activated carbon, and their impact on the design of CIL and CIP circuits. Minerals Engineering. 2011; 24(6): 484–494. doi: 10.1016/j.mineng.2011.03.021

[12] Soleimani M, Kaghazchi T. Adsorption of gold ions from industrial wastewater using activated carbon derived from hard shell of apricot stones—an agricultural waste. Bioresource Technology. 2008; 99(13): 5374–5383 doi:10.1016/j.biortech.2007.11.021

[13]   Espiell F, Roca A, Cruells M, Nunez C. Gold desorption from activated carbon with dilute NaOH/organic solvent mixtures. Hydrometallurgy. 1988; 19(3): 321–333. doi: 10.1016/0304-386X(88)90038-2

[14]   Muir DM, Hinchliffe W, Tsuchida N, Ruane M. Solvent elution of gold from CIP carbon. Hydrometallurgy. 1985; 14(1): 47–65. doi:10.1016/0304-386X(85)90005-2

[15]   Davidson RJ, Veronese V, Nkosi MV. The use of activated carbon for the recovery of gold and silver from gold-plant solutions. Journal of the South African Institute of Mining and Metallurgy. 1979; 281–297.

[16]   Conradie PJ, Johns MW, Fowles RJ. Elution and electrowinning of gold from gold-selective strong-base resins. Hydrometallurgy. 1995; 37(3): 349–366. doi: 10.1016/0304-386X(94)00032-X

[17]   Marsden JO, House CI. The chemistry of gold extraction, Society for Mining, Metallurgy and Exploration. Colorado; Littleton: 2006. p. 147–231.

[18]   Yannopoulos JC. Melting and refining of gold. In the extractive metallurgy of gold. New York, NY: Springer; 1991. p. 241–244. doi:10.1007/978-1-4684-8425-0_11

[19]   Duchao Z, Tianzu Y, Wei L, Weifeng L, Zhaofeng X. Electrorefining of a gold-bearing antimony alloy in alkaline xylitol solution. Hydrometallurgy. 2009; 99(3): 151–156. doi: 10.1016/j.hydromet.2009.07.013

[20]   Haque MR, Bradbury JH. Total cyanide determination of plants and foods using the picrate and acid hydrolysis methods. Food Chemistry. 2002; 77(1): 107–114. doi:10.1016/ S0308-8146(01)00313-2

[21]   Dubey SK, Holmes DS. Biological cyanide destruction mediated by microorganisms. World Journal of Microbiology and Biotechnology. 1995; 11(3): 257–265. doi:10.1007/ BF00367095

[22]   Kuyucak N, Akcil A. Cyanide and removal options from effluents in gold mining and metallurgical processes. Minerals Engineering. 2013; 50: 13–29. doi:10.1016/j.mineng. 2013.05.027

[23]   Zheng DA, Dzombak RG, Luthy B, Sawer W, Lazouska P, Tata, MF, Sebriski JR, Swartling RS, Drop SM, Flaherty JM. Evaluation and testing of analytical methods for cyanide species in municipal and industrial contaminated water. Environmental Science and Technology. 2003; 37: 107–115. doi:10.1021/es0258273

[24]   Ebbs S. Biological degradation of cyanide compounds. Current Opinion in Biotechnology. 2004; 15(3): 231–236. doi:10.1016/j.copbio.2004.03.006

[25]   Desai JD, Ramakrishna C. Microbial degradation of cyanides and its commercial application, Journal of Scientific and Industrial Research. 1998; 57: 441–453.

[26]   Randol International Limited. Water management & treatment for mining & metallurgical operations. Colorado; Golden: 1985. p. 2294–2700.

[27] Flynn CM, Haslem SM. Cyanide Chemistry: Precious Metals Processing and Waste Treatment. US Department of the Interior, Bureau of Mines: 1995.

[28] Young CA, Jordan TS. Cyanide remediation: current and past technologies. In: Erickson LE, Tillison DL, Grant SC, McDonald JP. Proceedings of 10th Conference on Hazardous Waste Research; 23–24 May 1995. Kansas State University, Manhattan, Kansas: p. 104–129.

[29] Botz MM. Overview of cyanide treatment methods, mining environmental management. London: Mining Journal Ltd; 2001. p. 28–30.

[30] Dash RR, Balomajumder C, Kumar A. Removal of cyanide from water and wastewater using granular activated carbon. Chemical Engineering Journal. 2009; 146: 408–413. doi: 10.1016/j.cej.2008.06.021

[31] Ripley EA, Redmann RE, Crowder AA. Environmental effects of mining. Florida: St.-Lucie Press; 1996. p 181–197.

[32] USEPA (Environ. Protection Agency U.S.) Drinking water criteria document for cyanide, Environment Criteria and Assessment Office. Cincinnati. 1985; EPA/600/X-84-192-1.

[33] Parga JR, Shukla SS, Carrillo-Pedroza FR. Destruction of cyanide waste solutions using chlorine dioxide, ozone and titania sol. Waste Management. 2003; 23: 183–191. doi: 10.1016/S0956-053X(02)00064-8

[34] Akcil A. Destruction of cyanide in gold mill effluents: biological versus chemical treatments. Biotechnology Advances. 2003; 21(6): 501–511. doi:10.1016/S0734-9750(03)00099-5

[35] Gurbuz F, Ciftci H, Akcil A. Biodegradation of cyanide containing effluents by Scenedesmus obliquus. Journal of HAZARDOUS materials. 2009; 162(1): 74–79. doi:10.1016/j.jhazmat.2008.05.008

[36] Luthy RG, Bruce Jr SG. Kinetics of reaction of cyanide and reduced sulfur species in aqueous solution. Environmental Science & Technology. 1979; 13(12): 1481–1487. doi: 10.1021/es60160a016

[37] Botz M, Mudder T, Akcil A. Cyanide treatment: physical, chemical and biological processes. Advances in Gold Ore Processing. 2005; 672–700.

[38] Karunakaran C, Abiramasundari G, Gomathisankar P, Manikandan G, Anandi V. Preparation and characterization of $ZnO–TiO_2$ nanocomposite for photocatalytic disinfection of bacteria and detoxification of cyanide under visible light. Materials Research Bulletin. 2011; 46(10): 1586–1592. doi:10.1016/j.materresbull.2011.06.019

[39] Moussavi G, Khosravi R. Removal of cyanide from wastewater by adsorption onto pistachio hull wastes: Parametric experiments, kinetics and equilibrium analysis.

Journal of Hazardous Materials. 2010; 183(1): 724–730. doi:10.1016/j.jhazmat. 2010.07.086

[40] Kitisa M, Karakayaa E, Yigita NO, Civelekoglua G, Akcilb A. Heterogeneous catalytic degradation of cyanide using copper-impregnated pumice and hydrogen peroxide. Water Research. 2005; 39: 1652–1662. doi:10.1016/j.watres.2005.01.027

[41] Depci T. Comparison of activated carbon and iron impregnated activated carbon derived from Gölbaşı lignite to remove cyanide from water. Chemical Engineering Journal. 2012; 181: 467–478. doi:10.1016/j.cej.2011.12.003

[42] Campos MG, Pereira P, Roseiro JC. Packed-bed reactor for the integrated biodegradation of cyanide and formamide by immobilised *Fusarium oxysporum* CCMI 876 and *Methylobacterium sp.* RXM CCMI 908. Enzyme and Microbial Technology. 2006; 38: 848–854. doi:10.1016/j.enzmictec.2005.08.008

[43] Kao CM, Liu JK, Lou HR, Lin CS, Chen SC. Biotransformation of cyanide to methane and ammonia by *Klebsiella oxytoca*. Chemosphere. 2003; 50: 1055–1061. doi:10.1016/ S0045-6535(02)00624-0

[44] Devuyst EA, Conard BR, Ettel VA. Pilot-plant operation of the Inco SO_2 air cyanide removal process. Canadian Mining Journal. 1982; 103(8): 27–30.

[45] Mudder TI, Botz M, Smith A. Chemistry and treatment of cyanidation wastes. London: Mining Journal Books; 2001. p. 327–333.

[46] Nelson GM, Kroeger EB, Arps PJ. Chemical and biological destruction of cyanide: comparative costs in a cold climate. Mineral Processing and Extractive Metallurgy Review. 1998; 19: 217–226. doi:10.1080/08827509608962441

[47] Kepa U, Stanczyk-Mazanek E, Stepniak L. The use of the advanced oxidation process in the ozone+ hydrogen peroxide system for the removal of cyanide from water. Desalination. 2008; 223(1): 187–193. doi:10.1016/j.desal.2007.01.215

[48] Carrillo-Pedroza FR, Nava-Alonso F, Uribe-Salas A. Cyanide oxidation by ozone in cyanidation tailings: Reaction kinetics. Minerals Engineering. 2000; 13(5): 541–548. doi: 10.1016/S0892-6875(00)00034-0

[49] Rowley WJ, Otto FD. Ozonation of cyanide with emphasis on gold mill wastewaters. The Canadian Journal of Chemical Engineering. 1980; 58(5): 646–653. doi:10.1002/cjce. 5450580516

[50] Soto H, Nava F, Leal J, Jara J. Regeneration of cyanide by ozone oxidation of thiocyanate in cyanidation tailings. Minerals Engineering. 1995; 8(3): 273–281. doi: 10.1016/0892-6875(94)00123-T

[51] Zeevalkink JA, Visser DC, Arnoldy P, Boelhouwer C. Mechanism and kinetics of cyanide ozonation in water. Water Research. 1980; 14(10): 1375–1385. doi: 10.1016/0043-1354(80)90001-9

[52] Castrantas HM, Manganaro JL, Rautiola CW, Carmichael J. Treatment of cyanides in effluents with Caro's acid. U.S. Patent 5,397,482, issued March 14, 1995.

[53] Adams MD. The removal of cyanide from aqueous solution by the use of ferrous sulphate. Journal of the South African Institute Mining and Metallurgy. 1992; 92: 17–25.

[54] Marsden J, House I. The Chemistry of gold extraction. New York, NY: Ellis Horwood; 1992. p. 160–170.

[55] Kenfield CF, Qin R, Semmens MJ, Cussler EL. Cyanide recovery across hollow fiber gas membranes. Environmental Science & Technology. 1988; 22(10): 1151–1155. doi: 10.1021/es00175a003

[56] Han B, Shen Z, Wickramasinghe SR. Cyanide removal from industrial wastewaters using gas membranes. Journal of Membrane Science. 2005; 257(1): 171–181. doi:10.1016/j.memsci.2004.06.064

[57] Marder L, Sulzbach GO, Bernardes AM, Ferreira JZ. Removal of cadmium and cyanide from aqueous solutions through electrodialysis. Journal of the Brazilian Chemical Society. 2003; 14(4): 610–615. doi:10.1590/S0103-50532003000400018

[58] Nicol MJ, Fleming CA, Paul R.L. The Chemistry of the extraction of Gold, In: G.C. Stanley (Ed.), Extraction metallurgy of gold in South Africa. Pretoria: SAIMM; 1987. p. 834–905.

[59] Kuhn AT. Electrolytic decomposition of cyanides, phenols and thiocyanates in effluent streams—a review. Journal of Chemical Technology and Biotechology. 1971; 21(2): 29–34. doi:10.1002/jctb.5020210201

[60] Goldblatt E. Recovery of cyanide from waste cyanide solution by ion exchange. Industrial and Engineering Chemistry. 1959; 51, 241–246. doi:10.1021/ie51394a022

[61] Avery NL, Fries W. Selective removal of cyanide from industrial waste effluents with ion-exchange. Industrial Engineering Chemistry Product and Research Development. 1975; 14: 102–104. doi:10.1021/i360054a009

[62] Akser M, Wan RY, Miller JD. Gold adsorption from alkaline aurocyanide solution by neutral polymeric adsorbents, Solvent Extraction and Ion Exchange. 1986; 4: 531–546. doi:10.1080/07366298608917880

[63] Fleming CA, Cromberge G. The extraction of gold from cyanide solutions by strong and weak-base anion-exchange resins. Journal of Africa Institute of Mining and Metallurgy. 1984; 84: 125–138.

[64] Strobel G.A. Cyanide utilization in soil. Soil Science. 1967; 103: 299–302.

[65] Saxena S, Prasad M, Amritphale SS, Chandra N. Adsorption of cyanide from aqueous solutions at pyrophyllite surface. Separation and Purification Technology. 2001; 24(1): 263–270. doi:10.1016/S1383-5866(01)00131-9

[66] Honda S, Kondo G. Treatment of wastewater containing cyanide using activated charcoal. Os. Kogyo Gij. Shik. Koho. 1967; 18: 367.

[67] Dash RR, Majumder CB, Kumar A. Treatment of metal cyanide bearing wastewater by simultaneous adsorption biodegradation (SAB). Journal of Hazardous Materials. 2008; 152: 387–396. doi:10.1016/j.jhazmat.2007.07.009

[68] Adhoum N, Monser L. Removal of cyanide from aqueous solution using impregnated activated carbon. Chemical Engineering and Processing: Process Intensification. 2002; 4:117–121. doi:10.1016/S0255-2701(00)00156-2

[69] Deveci H, Yazıcı EY, Alp I, Uslu T. Removal of cyanide from aqueous solutions by plain and metal-impregnated granular activated carbons. International Journal of Mineral Processing. 2006; 79: 198–208. doi:10.1016/j.minpro.2006.03.002

[70] Crini G. Non-conventional low-cost adsorbents for dye removal: a review. Bioresource Technology. 2006; 60: 67–75. doi:10.1016/j.biortech.2005.05.001

[71] Mosher JB, Figueroa L. Biological oxidation of cyanide: a viable treatment option for the minerals processing industry. Minerals Engineering. 1996; 9(5): 573–581. doi: 10.1016/0892-6875(96)00044-1

[72] Luque-Almagro VM, Huertas MJ, Martínez-Luque M, Moreno-Vivián C, Roldán, MD, García-Gil LJ, Castillo F, Blasco R. Bacterial degradation of cyanide and its metal complexes under alkaline conditions. Applied and Environmental Microbiology. 2005; 71(2): 940–947. doi:10.1128/AEM.71.2.940-947.2005

[73] Gurbuz F, Ciftci H, Akcil A, Karahan AG. Microbial detoxification of cyanide solutions: a new biotechnological approach using algae. Hydrometallurgy. 2004; 72(1): 167–176. doi:10.1016/j.hydromet.2003.10.004

[74] Akcil A, Mudder T. Microbial destruction of cyanide wastes in gold mining: process review. Biotechnology Letters. 2003; 25: 445–450. doi:10.1023/A:1022608213814

[75] Kuyucak N. The Role of microorganisms in mining: generation of acid rock drainage, its mitigation and treatment. European Journal of Mineral Processing and Environmental Protection. 2002; 2(3): 179–196.

[76] Baxter J, Cummings SP. The current and future applications of microorganism in the bioremediation of cyanide contamination. Antonie van Leeuwenhoek, 2006; 90: 1–17. doi:10.1007/s10482-006-9057-y

[77] Dumestre A, Chone T, Portal J, Berthelin J. Cyanide degradation under alkaline conditions by a strain of *Fusarium solani* isolated from contaminated soils. Applied and Environmental Microbiology. 1997; 63: 2729–2734.

[78] Dursun AY, Aksu Z. Biodegradation kinetics of ferrous (II) cyanide complex ions by immobilized *Pseudomonas fluorescens* in a packed bed column reactor. Process Biochemistry. 2000; 35: 615–622. doi:10.1016/S0032-9592(99)00110-7

[79] Babu GRV, Wolfram JH, Chapatwala KD. Conversion of sodium cyanide to carbon dioxide and ammonia by immobilized cells of *Pseudomonas Putida*. Journal of Industrial Microbiology. 1992; 9: 235–238. doi:10.1007/BF01569629

[80] Barclay M, Hart A, Knowles CJ, Meeussen JCL, Tett VA. Biodegradation of metal cyanides by mixed and pure cultures of fungi. Enzyme and Microbial Technology. 1998; 22: 223–231. doi:10.1016/S0141-0229(97)00171-3

[81] Kao CM, Chen KF, Liu JK, Chou SM, Chen SC. Enzymatic degradation of nitriles by *Klebsiella oxytoca*. Applied Microbiology and Biotechnology. 2006; 71: 228–233. doi: 10.1007/s00253-005-0129-0

[82] Randol International Limited. Phase IV – Innovations in gold and silver recovery. Colorado: Golden; 1992. Chapter 54.

[83] Cushnie G. Pollution prevention and control technologies for plating operations, Wastewater Treatment. 2nd ed. London; Amazon: 2009. p. 287–297.

[84] Saarela K, Kuokkanen T. Alternative disposal methods for wastewater containing cyanide: analytical methods for new electrolysis technology developed for total treatment of wastewater containing gold or silver cyanide. In: Pongracz, E. (Ed.), Proceedings of the Waste Minimization and Resource Use Optimization Conference; 2004. p. 107–121 University of Oulu, Oulu. Finland.

2

A Comparative Study of Modified and Unmodified Algae (*Pediastrum boryanum*) for Removal of Lead, Cadmium and Copper in Contaminated Water

John Okapes Joseph, Isaac W. Mwangi,

Sauda Swaleh, Ruth N. Wanjau, Manohar Ram and

Jane Catherine Ngila

Additional information is available at the end of the chapter

Abstract

The presence of heavy metals in water is of concern due to the risk toxicity. Thus there is need for their removal for the safety of consumers. Methods applied for removal of heavy metals include adsorption, membrane filtration and co-precipitation. However, studies have revealed adsorption is highly effective technique. Most adsorbents are expensive or require extensive processing before use and hence need to explore for possible sources of inexpensive adsorbents. This research work investigated the use an algal biomass (*pediastrum boryanum*) as an adsorbent for removal of Lead, Cadmium and Copper in waste water in its raw and modified forms. The samples were characterized with FTIR and was confirmed a successful modification with tetramethylethlynedia-mine (TMEDA). Sorption parameters were optimized and the material was finally applied on real water samples. It was found that the sorption was best at lower pH values (4.2-6.8). Sorption kinetics was very high as more that 90% of the metals were removed from the solution within 30 minutes. The adsorption of copper fitted into the Langmuir adsorption isotherm indicating a monolayer binding mechanism. Cadmium and lead fitted best the Freundlich adsorption mechanism. Sorption of lead and cadmium was of pseudo-second order kinetics, confirming a multisite interaction whereas copper was pseudo-first order indicating a single site adsorption. The adsorption capacity did not improve upon modification but the stability of the material was improved and secondary pollution of leaching colour was alleviated. This implies that the modified material is suitable for application on the removal of metals from water.

Keywords: modified algae, tetramethylethylenediamine, sorption, heavy metals, con-taminated water

---

1. Introduction

The release of wastewater into the environment poses a great problem worldwide due to enhancement and mobilization of toxic trace metals due to solubilization [1, 2]. This is enabled by the presence of functional groups capable of forming metal complexes with the metals [3–5]. Unlike organic pollutants which are susceptible to degradation, metal ions remain in the environment available to cause pollution [6]. This makes the presence of heavy metal in the environment a major concern due to their toxicity to various life forms. Even when most metals in the environment are in trace levels or masked by matrices, the presence of wastewater problem exacerbates the toxic nature of heavy metals. The net result is scarcity and insufficient supply of safe water, hence the quality of life. As such, methods for the removal of such contaminants need to be explored to mitigate the effects of metal pollution.

Conventional methods for the removal of metals such as precipitation, coagulation, evaporation and membrane filtration are expensive and not effective when the concentrations are in trace levels ranging from 1 mg l^{-1} to 20 mg l^{-1} [7, 8]. Due o such limitations, a need therefore arises for the development of cost-effective methods to remove heavy metals in waste purification processes. The presence of functional groups within the structure of sorbents from plant origin has received increasing attention for the removal and recovery of heavy metals in aqueous media [9]. However, they have been found to leach organic matter in the water during the treatment process. This has resulted in treating the sorbents first, before applying them for the water treatment activity.

A solution to this was achieved by modification of the raw biomaterial using tetramethyle-thylenediamine [10]. This work reports on the modification of algae and its subsequent application for the removal of some selected heavy metals through a biosorption process. The effects of modification on secondary pollution and adsorption parameters were also investigated. This was to obtain the information that will contribute to determining adsorption capacity, sorption mechanism and kinetics with a view to apply the material at a point of use. This was intended to contribute to knowledge for social-economic development, to address the availability of clean water to the rural communities who source their water from rivers, dams, boreholes and shallow wells whose water quality is not known. The adsorbent is intended to be applied for purification of water by removal of heavy metals for domestic consumption at a small scale with a view to offer a cheap solution to metal-related toxicity. This is a simple and sustainable water management approach.

Methods applied for separation and preconcentration techniques include adsorption, membrane filtration, cloud-point extraction, solvent extraction and co-precipitation [11]. Studies by Marshall [12] have revealed adsorption by the use of activated carbon to be highly effective for the removal of heavy metals from wastewaters. Despite its extensive use in water and waste-water treatment industries, activated carbon remains an expensive material [13]. In view of this, the need for safe and economical methods for the elimination of heavy metals from polluted and contaminated water has to be explored.

This has necessitated researchers to develop methods to mitigate the effects of heavy metals in water. Low-cost agricultural waste by-products such as sugarcane, bagasse, rice husks, sawdust and coconut husk, oil palm shell, neem bark and maize tassels have been studied and applied for the removal of heavy metals from wastewater. The cost of such materials is an important parameter for comparing the sorbent materials. Such agricultural wastes are abundant, require little processing and therefore have a potential to be applied as low-cost sorbents that are environmentally friendly [14]. But these materials suffered a setback due to leaching of dissolved organic matter in water during the treating process. To overcome such cases of secondary pollution, there is need to explore sorbents for their feasibility for the removal of heavy metals from wastewater. This study investigated the use of algal biomass which is an aquatic plant for the remediation process. Algae does not normally leach colour in the water but only contribute in the water oxidation process and interaction with metals.

Due to those qualities of algae, a solution experienced while using other biosorbents was expected to be solved. The algal biomaterial was modified with tetramethylethylenediamine to form a resin material with suitable functional group to complex with metal ions and remove them from water. The resulting solid material was capable of interacting with metals in water and attracting them to its surface, hence removing them, and was found to be regeneratable and did not leach soluble organic compounds in the treated water.

2. Materials and methods

2.1. Research design

The focus of our study was to synthesize a sorbent by anchoring functional groups capable of interacting with metal ions and removing them from aqueous media. The protocol in prepa-ration was to use non-toxic and environmentally friendly materials. The study was carried out in several parts. This comprised of sampling, synthesis, characterization of the modified material, optimization of removal parameters and then its subsequent application for the removal of fluorides from both synthetic and environmental water samples.

2.2. Chemicals and reagents

All the solutions were prepared in double-distilled water and the reagents were of analytical grade. Metal standard stock solution of 1000 mg/L was prepared by dissolving 1.00 g of the respective metal in one litre of 0.1 M sodium acetate. It was from this solution that subsequent working solutions were prepared from. Separate solutions of 0.01 M nitric acid and 0.01 M sodium hydroxide were prepared and used to adjust the pH of the working solutions to the desired value. The above chemicals and tetramethylethylenediamine (TMEDA) were supplied by Kobian Kenya Ltd. which is Sigma-Aldrich's outlet in Kenya.

2.3. Instrumentation

The modified and unmodified algae were characterized using a Fourier transform infrared (FTIR) spectrometer (FTIR-8400, Shimadzu Tokyo, Japan) to establish the functional groups present. The concentration of metal pollutants in the water samples was determined using atomic absorption spectroscopy (AAS) (Buck, model 210 VGP) set at the optimum operating conditions and wavelength of each respective metal. All pH measurements were done using a calibrated (Jenway 3505) pH metre equipped with a standard calomel electrode (SCE). A constant shaker model CFC 3018 with a water bath was used to shake the samples at the required shaking speeds.

2.4. Sampling and pretreatment of the algae

Samples of the algae *Pediastrum boryanum* were collected in Molo, Nakuru County, Kenya. The algal material was cleaned using tap water, dried, ground into powder and then stored in clean plastic bottles. The dried powdered algae were used as a sorbent and for the modification process.

2.5. Modification of the dried algae

The algae were modified by anchoring tetramethylethylenediamine onto its chemical structure to improve its chelating property [15]. The modification procedure involved chlorination of the biomaterial first and then condensing the resultant with the amino compound. A sample of the algae (30 g) was chlorinated using thionyl chloride ($SOCl_2$) 100 ml, and the mixture heated under reflux at a temperature of 80°C for 4 h with continuous stirring. The chlorinated biomaterial was then washed with 100 ml of distilled water. The solution was filtered using Whatman No. 1 filter paper and dried in an oven set at 60°C. The dry chlorinated sample was then treated with 25.0 ml of tetramethylethylenediamine and refluxed for 3 h to anchor the tetramethylethylenediamine structure into the algal biomass.

2.6. Batch sorption experiments

Sorption studies were carried out on a mechanical reciprocating shaker (SKZ-1 NO. 1007827, India) using plastic screw cap bottles. Batch experiments were conducted to investigate the effects of pH, adsorbent dosage, adsorbate concentration and contact time on the adsorption of Pb^{2+}, Cd^{2+} and Cu^{2+} on the modified and unmodified algae. The pH of the model test solutions containing a known concentration of the metal ion was adjusted to values between 3 and 7. A known weight of the sorbent (0.03 g) was added to each of the solutions and then allowed to equilibrate giving sufficient time for sorption. The resulting mixture was filtered through Whatman No. 42 filter paper, and the metal ions in the filtrate were determined by atomic adsorption spectrophotometry (UNICAM 919).

2.7. Optimization of sorption parameters

2.7.1. Effect of pH on sorption

2.7.1.1. Calibration of the pH metre

The pH metre was calibrated using special buffer tablets for pH 3.0, 5.0, 7.0 and 9.0. Each tablet was dissolved separately in 100 ml of distilled water and then used. The electrode of the pH metre was conditioned with saturated potassium chloride overnight to wet the membrane and make it more sensitive. It was later calibrated with the buffer solutions. This procedure was undertaken before any pH measurements were made [16].

2.7.1.2. Effect of pH of adsorption

The effect of pH of the adsorbate on the adsorption of the metal ions by both modified and unmodified algae was investigated by mixing 0.2 g of the sorbent material with 50 ml of 10 ppm model solution buffered at different pH environments. The pH was brought to the desired values (3–10) by adding drops either 0.01 M nitric acid or 0.01 M sodium hydroxide. The resulting mixture was allowed to equilibrate for 120 min. The resulting mixture was then filtered through Whatman No. 1 and the concentration of the metal ions in the filtrate determined by atomic absorption spectroscopy.

2.7.2. Effect of contact time on sorption

The effect of contact time on sorption of lead (II), cadmium (II) and copper (II) by modified and unmodified algae was done by taking a sample, 0.2 g of the sorbent (modified and unmodified algae) into the plastic bottles and 50 ml of the adsorbate of concentration 10 ppm added. The mixture was buffered to the optimum pH value for each metal and agitated at predetermined time intervals of 2–150 min. The samples were then removed from the shaker and the solutions filtered, and the metal ion concentration in the filtrate was determined.

2.7.3. Effect of initial metal ion concentration on sorption

The extent to which metal ions are adsorbed as a function of the initial ion concentration was investigated by mixing 0.2 g of finely ground modified and unmodified algae separately with 50 ml of varying concentrations of the test solutions, buffered at the optimal pH value for each respective metal. The respective mixtures were allowed to equilibrate for a sufficient duration and then filtered and the concentration of the metal ions in the filtrate was determined.

2.7.4. Effect of sorbent dose on percentage metal removal

The effect of sorbent dose was investigated by agitating 50 ml (10 μg ml^{-1}) of the adsorbate solutions of lead (II), cadmium (II) and copper (II) with various dosages of the sorbents. 0.1, 0.5, 1.0, 1.5 and 2.0 g of modified and unmodified algae were used. The solutions of the adsorbate were buffered to the optimum pH of each respective metal ion under investigation.

The solutions were for 2 h with the temperature set at 25°C. The resulting mixtures were filtered and the concentration of the residual metal ions determined.

2.7.5. Determination of adsorption capacity of modified and unmodified algae

The adsorption capacity was determined by mixing 0.2 g of finely ground sorbent material with 50 ml of varying concentrations of the test metal solution (concentrations 10–250 ppm) buffered at the optimum pH for each respective metal. The mixtures were agitated for 30 min and then filtered, and the concentrations of metal ions were determined.

2.7.6. Adsorption models

The experimental data on metal sorption were also analyzed using adsorption models so as to establish the sorption kinetics and mechanism.

2.7.7. The kinetics of adsorption

To determine the necessary time, different solutions of for adsorption lead (II), cadmium (II) and copper (II) 20 ml containing 10 µg ml^{-1} of the adsorbate were introduced in different sets of plastic bottles containing 0.2 g of the adsorbent, and the pH sets an optimal value for each metal. The mixtures were then introduced in the shaker temperature of 25°C and equilibrated at different time intervals of 2, 5, 10, 20, 30, 60, 90 and 120 min. They were then filtered, and the filtrate was analyzed for adsorbate concentration. The data obtained was treated with Lagergren's [17] pseudo-first-order and Ho et al.'s [18] pseudo-second-order equations to determine molecularity of the adsorption.

The Lagergren first-order and Ho's second-order kinetic models are expressed as shown in equations 5.1 and 5.2, respectively:

$$\ln(C_o - C_t) = Kt + A \tag{1}$$

$$\frac{1}{q_e} = Kt + A \tag{2}$$

where C_o is the adsorption per unit mass of adsorbent at equilibrium, K is the adsorption rate constant, A is intercept and C_t is the concentration at time t.

2.7.8. Adsorption isotherms

The experimental data for the effect of metal ion concentration obtained was treated with the Freundlich and Langmuir isotherm models to obtain the adsorption mechanism.

2.7.8.1. Langmuir isotherm

For molecules in contact with a solid surface at a fixed temperature, the Langmuir isotherm, developed by Irving Langmuir in 1918, describes the partitioning between gas phase and adsorbed species as a function of applied pressure [19]. Langmuir adsorption isotherm is the widely used isotherm for modeling of adsorption data [20]. Langmuir considered adsorption of an ideal gas on an ideal surface. It is based on the assumption that adsorption can only occur at fixed sites and only hold on one adsorbate molecule (monolayer). All sites are equivalent with no interaction between adsorbed molecules, and the sites are independent as reported by Langmuir [19]. The Langmuir equation was derived from Gibbs approach which takes the form shown in equation 5.3 [19, 21]:

$$q_e = \frac{K_L C_e}{1 + a_L C_e} \tag{3}$$

where C_e is the equilibrium concentration, K_L is the equilibrium constant, q_e is the metal concentration on the sorbent phase at equilibrium in mg g^{-1} and a_L is a Langmuir constant. Eq. (3) can be linearized and often referred to as linearized Langmuir equation as shown in equation 5.4:

$$\frac{C_e}{q_e} = \frac{1}{K_L} + \frac{a_L C_e}{K_L} \tag{4}$$

The experimental data was applied on the equation above, and a plot of $\frac{C_e}{q_e}$ against C_e gave a linear regression. This indicates that the adsorption prescribes to the Langmuir model, where the gradient $\left(\frac{a_L}{K_L}\right)$ is the theoretical saturation capacity (units in mg g^{-1}) and the intercept is $\frac{1}{K_L}$ [19, 22, 23].

2.7.8.2. Freundlich isotherm

Freundlich isotherm is an empirical equation based on heterogeneous surface [23]. This is a multi-site adsorption isotherm for heterogeneous surfaces and has a general form as shown in Eq. (5):

$$q_e = K_F C_e^{bF} \tag{5}$$

The equation was linearized by taking logarithms and then applied to determine if the systems are heterogeneous with highly interactive species [24]:

$$\ln q_e = \ln K_F + b_F \ln C_e \tag{6}$$

where q_e and C_e have the same meaning as in Eq. (3); the numerical value of K_F presents adsorption capacity, and b_F indicates the energetic heterogeneity of adsorption sites [24]. From the data, if a plot of **ln** q_e versus **ln** C_e gave a straight line, it indicates that the adsorption prescribes to the Freundlich model.

3. Results and discussion

3.1. Introduction

The algal material was modified with an amino compound to improve its thermal stability [9]. The resulting product obtained was a water-insoluble solid material. The raw and the modified products were characterized using FTIR to obtain the functional groups present; sorption parameters were established and then used for adsorption experiments in both synthetic solutions and real water samples.

3.2. FTIR analysis of modified and unmodified algae

The modified and unmodified materials were characterized with FTIR, and the resulting spectra are presented in **Figure 1**.

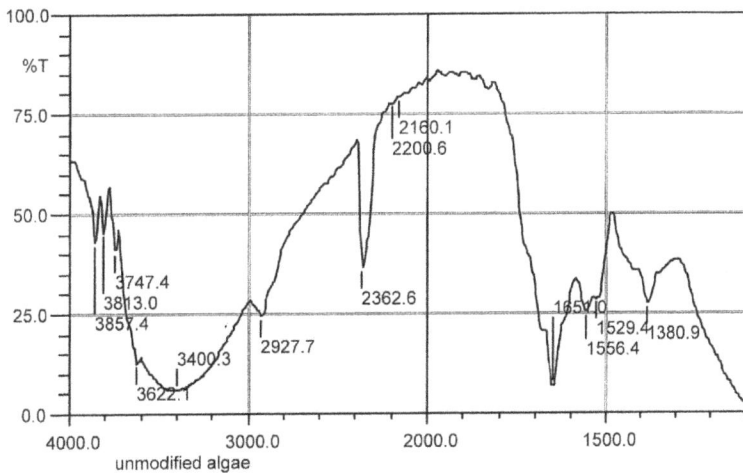

Figure 1. FTIR spectrum of unmodified algae.

The results show the presence of many functional groups capable of metal sorption. The broad and strong band at 3400.3 cm^{-1} could be attributed to either —OH or —NH group [25]. The band at 2927.72 cm^{-1} was assigned to C—H stretches, while the band at 1651.0 cm^{-1} was assigned to stretching —OH, C=O or N=C [25] .The band at 1380.9 cm^{-1} confirms the presence of an amide group of an amide or sulphamide group [26, 27]. The material was modified with tetramethy-lethylenediamine, and results obtained are presented in **Figure 2**.

Upon modification, the band at 3400.3 cm^{-1} that was attributed to the $-NH$ of an amide shifts to a lower value of 3394.5 cm^{-1}. The intensity of the band also decreases. This can be attributed to suppression influenced by the carbon atoms from anchored tetramethylethylenediamine functional groups. Similar observations were reported by Schluter [28] as he studied the treatment of poly(1.1.1)propellane with lithium organic initiators and then investigated its rigidity. Mwangi and Ngila [29] also recorded the same observation when studying the removal of heavy metals from contaminated water using ethylenediamine-modified green seaweed. New bands are also seen to appear at 1454.2 cm^{-1} and 2597 cm^{-1}. The band at 1454.2 cm^{-1} can be attributed to an NO$_2$ group and that at 2597 cm^{-1} can be attributed to an additional $-OH$ groups from a phenolic compound. Both the parent and the modified material were applied for sorption experiments.

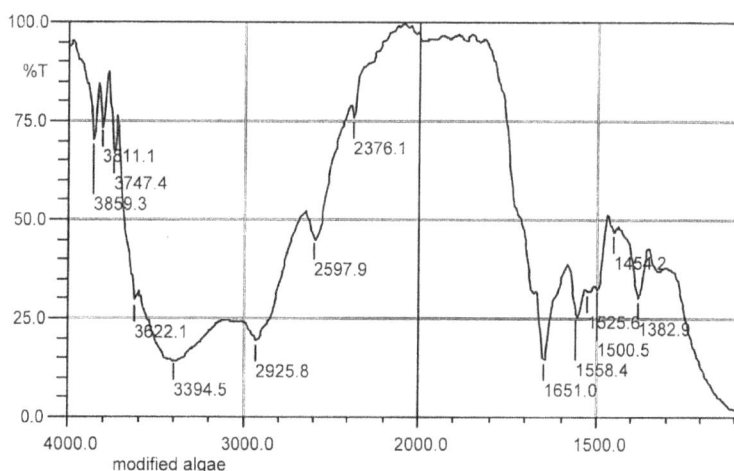

Figure 2. FTIR spectrum of modified algae.

3.3. Effect of pH on sorption

The adsorption of metal ions into the biosorbent is dependent on pH of the solution. pH affects the biosorbent surface charge and degree of ionization. The sorbent also has nitrogen atoms (with a lone pair of electrons) which can be influenced by the pH of the medium. The effect of pH on sorption of lead, cadmium and copper ions are represented in **Figure 3**.

The pH of the solution influences the chemistry of the metal binding sites and the behaviour of the metal itself in solution. The results show that the maximum adsorption for lead was found to occur at a pH of 3.5 by the unmodified sorbent and at a pH of 7.0 for this same metal by the modified adsorbent. There was an increase in the amount adsorbed as the pH increased from 3.5 to 7.0. Beyond this, as the pH increases, the amount adsorbed decreases. Similar results have been reported for other biosorbents [29]. Singh and co-workers (2006) also reported that the highest percentage of lead (II) ions was adsorbed by phosphatic clay at a pH of 5.0 [30]. Similar results were reported by Matheickal et al. [26] when they studied the biosorption of lead by marine alga *Ecklonia radiata*. Benhima et al. [31] observed that there was an increase in lead (II) ion uptake by inert organic matter (IOM) as the pH increased from 2.0 to 6.0. This is in agreement with the observed results for lead (II) ions.

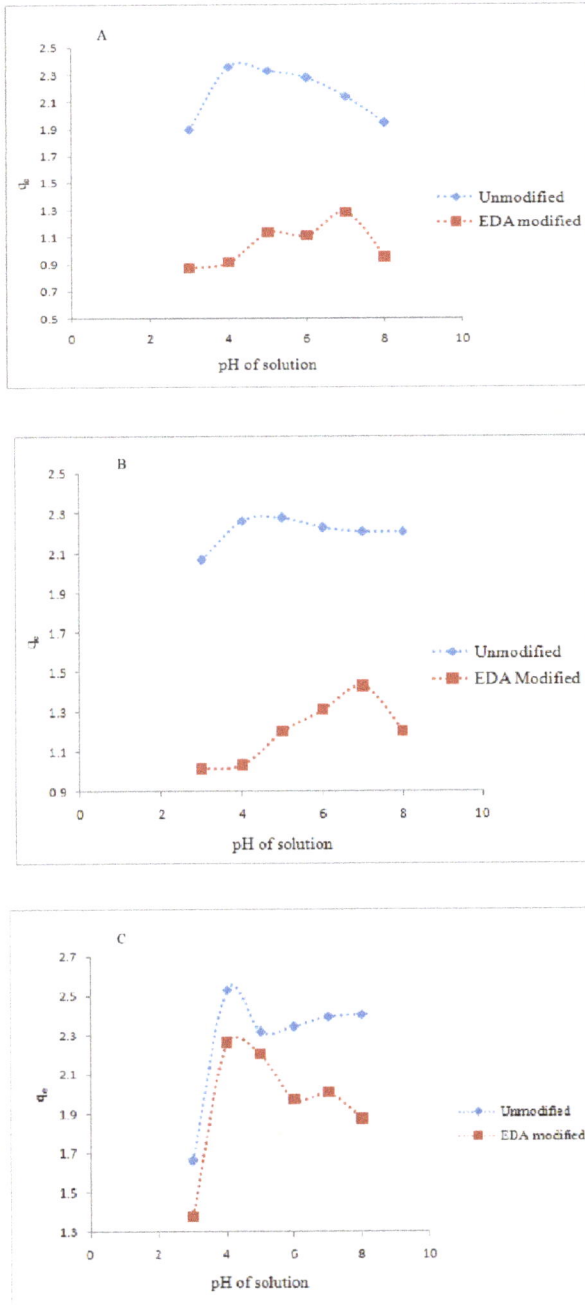

Figure 3. Effect of pH on adsorption of lead, cadmium and copper ions—A, B and C, respectively.

At low pH, the biomass surface would be completely covered with hydrogen ions. H^+ lead (II) ions cannot compete effectively for the adsorption sites. This can be attributed to the fact that protons are strongly competing due to their high concentration. Godhane et al. [32] reported that the minimal sorption obtained at low pH may be due to high mobility of protons and partly due to the fact that the solution pH influences the sorbent surface charge.

The unmodified biosorbent has a maximum adsorption for cadmium at a pH of 5.2, while the modified form was at a pH of 6.7. Similar results were obtained by Singh and co-workers [30] when they investigated the adsorption of cadmium using phosphatic clay. They observed maximum adsorption at a pH of 5.4.

Copper unlike the other metals has a maximum adsorption at a pH of 4.2 for both the modified and unmodified sorbents. This can be explained by the small size of copper giving it a high polarizing power on electrons of the adsorbent [33].

The sorbent has nitrogen atoms (with alone pair of electrons) as well as other functional groups, all of which may be influenced by pH. At low pH, the adsorbent is positively charged because the pH is lower than the isoelectric point or point of zero charge (PZC), i.e. pH < PZC. At such low pH range, adsorption is poor due to the charge on the adsorbent [34]. At high pH (pH > PZC), the adsorbent is negatively charged contributing to the high adsorption [26]. This arises from the fact that when the metal is in solution, it is positively charged and will be attracted to the surface of the negatively charged adsorbent at pH > PZC favouring adsorption. At pH > 7, there is metal hydrolysis leading to precipitation due to formation of hydroxyl metal ions [35].

3.4. Effect of contact time on sorption

The dependence of the adsorption process on the residence time of the adsorbate at the solid solution interphase was studied using batch sorption experiments for both the modified and unmodified algae. The solutions were set at optimal pH values of each metal, and the results of time-dependent adsorption obtained are shown in **Figure 4**.

It was observed that the general uptake rate was fast as over 90% of the adsorption which took place within the first 30 min for all the three metals after which a steady adsorption rate was realized (**Figure 3**). Keskinkan et al. [36] reported similar findings while studying the biosorption of lead (II) ions using aquatic *Ceratophyllum demersum*. Also, Yang and Volesky [37] made similar observations while studying the biosorption of cadmium (II) ions using dead brown *Sargassum fluitans*. The initial rapid uptake may be due to the physical adsorption or ion exchange at the cell surface and the subsequent slower phase due to the other mechanisms such as complexation, micro-precipitation or saturation at the binding sites [38]. Generally, there was a decrease in adsorption for the modified sorbent with the decrease more pronounced for cadmium (II) ions. Different functional groups with different affinities for the metal ions are usually present on the biomass surface; these are significantly altered by modification. New groups are also introduced which affect the binding ability of the sorbent.

An example of a functional group that was affected is the –NH group which appeared at 3400.3 cm^{-1} and after modification shifted to a lower value of 3394.5 cm^{-1}.

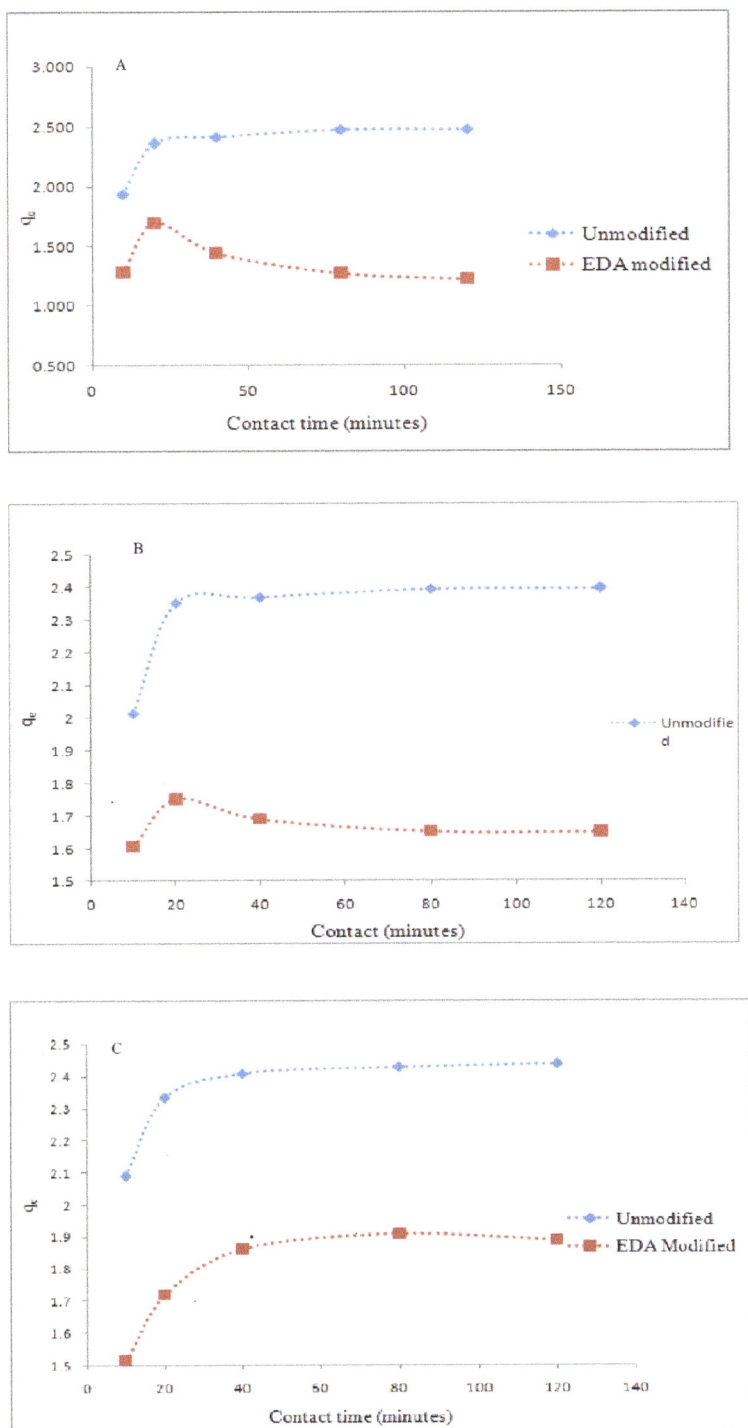

Figure 4. Effect of contact time on sorption of lead, cadmium and copper—A, B and C, respectively.

3.5. Effect of sorbent dosage on sorption

Results presented in **Figure 5** show the effect of varying the mass of the sorbent on the adsorption of the metal ions. The experiment was performed while the solutions were buffered at optimal pH value of each respective metal ion.

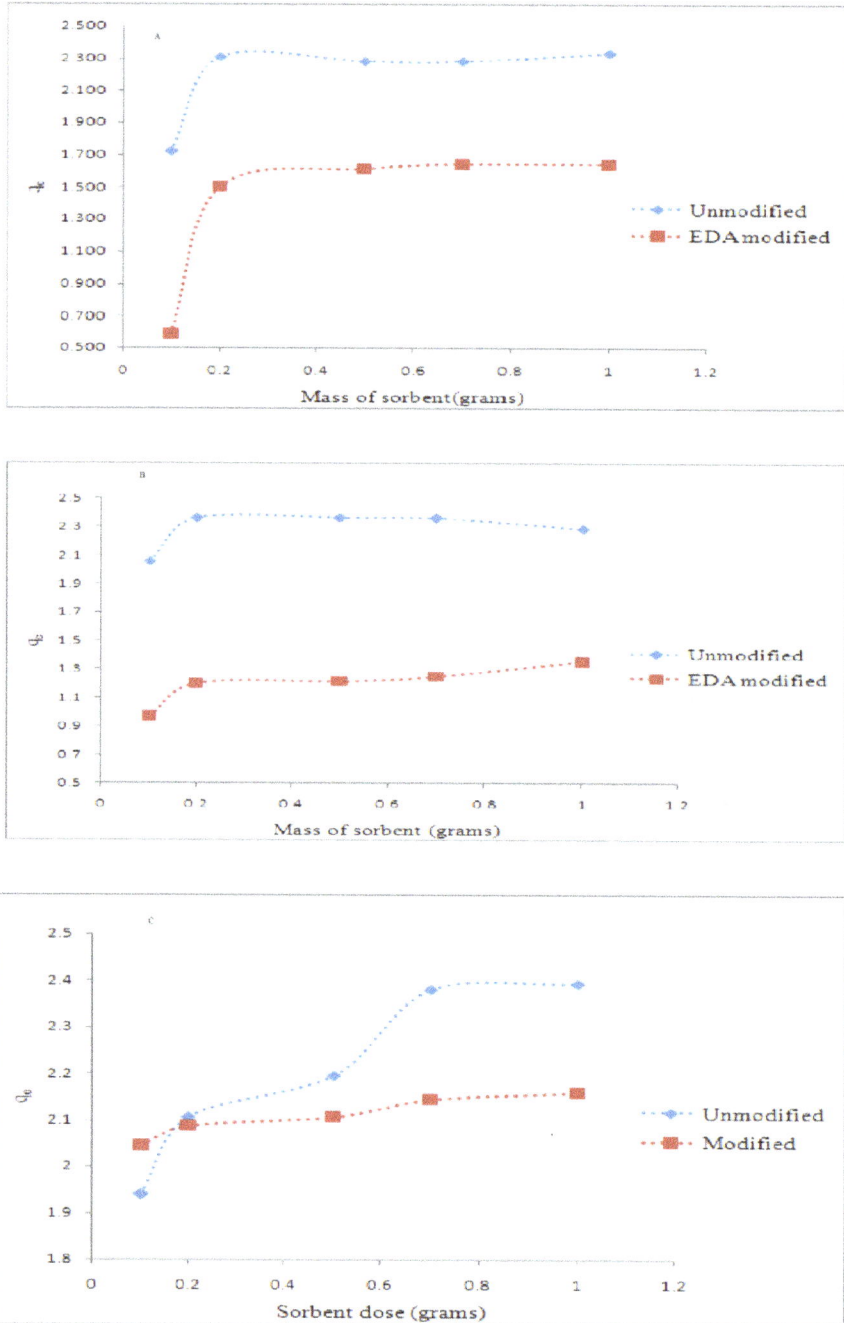

Figure 5. Effect of sorbent dose on sorption of lead, cadmium and copper — A, B and C, respectively.

The general observation is that the amount adsorbed metal ions increased with increase in sorbate dose. This can be attributed to the fact that as the concentration of the adsorbent increases, more adsorption sites are available due to increased surface area. More metal ions can therefore occupy those available active sites [39]. It was also noted that for both the unmodified adsorbates, copper was adsorbed more than both lead and cadmium. This could be attributed to the fact that the affinity to the binding sites is related to the ionic sizes of the respective metals, hence the polarizing power of copper being responsible for the observation. Maximum adsorption by the unmodified sorbate occurred at a mass of about 0.2 g for both lead and cadmium. Mwangi and Ngila [29] reported a similar observation while studying the biosorption of lead and cadmium using green seaweed, *Caulerpa serrulata*. Higher metal ion uptake at low sorbate mass concentrations has been attributed to an increased metal to biosorbent ratio which decreases upon an increase in sorbate mass concentration [40]. The modified sorbate showed maximum adsorption at a mass of 0.7 g for these two metals. For copper, maximum adsorption occurred at a sorbate mass of about 0.7 g for both modified and unmodified algae *Pediastrum boryanum*. A plateau is reached after a certain mass for all the metal ions. The unmodified sorbent material removed more metal ions than the modified material. This decrease in the adsorption capacity could be attributed by the fact that the functional groups in the parent material responsible for binding the metals had higher complexation ability than the anchored group [41]. A similar observation was reported by Drake and co-workers as they studied the sorption of chromium by *Datura inoxia* biomaterial [42].

3.6. Effect of initial ion concentration for determination of adsorption capacity

In order to investigate the adsorption capacity of both modified and unmodified algae, 50 ml solutions of concentrations ranging from 10 µg l^{-1} to 100 µg l^{-1} were added to 0.2 g of the biosorbent. **Figure 5** shows the results that were obtained.

A linear pattern of metal uptake was observed followed by some plateau for all metals. This may be attributed to the saturation of the binding sites as the concentration of the ion increases resulting to a steady state. The saturation is more pronounced at very high concentrations [31]. This is attributed to the fact that concentration is the driving force for metal ions to occupy the available binding sites [38]. It was observed that there was no significant difference in the sorption of copper and cadmium by the parent and modified material (see **Figure 6**).

The experimental data above was fitted into the Langmuir and Freundlich adsorption isotherm represented by equations (3) and (4) in Chapter 3 to determine the adsorption capacity for the metal ions using both modified and unmodified algal materials. **Table 1** is a summary of the results that were obtained.

Table 1 shows that the adsorption for both cadmium and lead fitted well with the Freundlich model, while copper fitted well with the Langmuir model. Similar results were obtained by Ng and co-workers [22] as they studied the adsorption of metals on cross-linked seaweed. This indicates that the adsorption process for lead and cadmium is a multi-site or physical sorption as a result of weak Van der Waals forces.

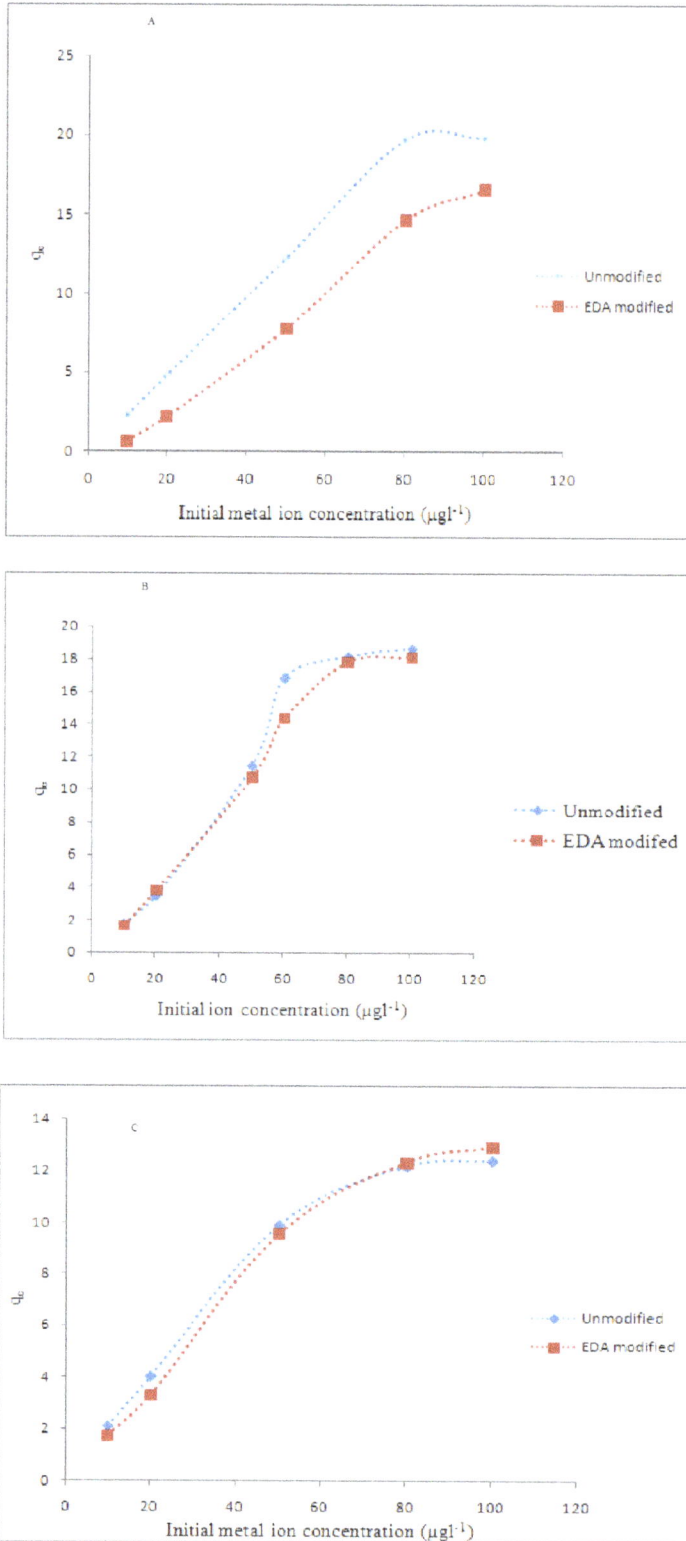

Figure 6. Effect of initial ion concentration on sorption of lead, cadmium and copper ions—A, B and C, respectively.

Metal	Langmuir		Freundlich		Comment
	R^2	Adsorption capacity	R^2	K_F/mg g^{-1}	
Copper					
Unmodified	0.974	0.059	0.605	0.948	Langmuir
Modified	0.512	0.001	0.502	–	Langmuir
Cadmium					
Unmodified	0.043	0.060	0.251	1.137	Freundlich
Modified	0.941	0.276	0.993	2.843	Freundlich
Lead					
Unmodified	0.570	0.021	0.929	0.791	Freundlich
Modified	0.560	0.112	0.906	1.695	Freundlich

Table 1. Results for Langmuir and Freundlich models.

Copper on the other hand fitted well with the Langmuir model with R^2 values of 0.974 for the unmodified and 0.512 for the modified algal material. These values show that the data has a strong correlation and therefore the sorption mechanism can be prescribed to the Langmuir model. The adsorption capacities were found to be 0.059 and 0.001 mg g^{-1} for copper, by the unmodified and modified materials, respectively. Sorption of cadmium and lead prescribed to the Freundlich model with a sorption of 1.137 and 2.843 and 0.791 and 1.695, respectively, by the unmodified and modified adsorbent materials in the same order.

3.7. Kinetics of adsorption of the metal ions

Lagergren's first-order and Ho's second-order kinetics were applied to the data that was obtained [17, 18]. This was used to determine the molecularity of the adsorption and the rate controlling step. A summary of the results that were obtained for all the three metals is shown in **Table 2**.

Metal	Lagergren	Ho	Comment
	R^2	R^2	
Lead			
Modified	0.936	0.954	Pseudo-second order
Unmodified	0.846	0.849	Pseudo-second order
Cadmium			
Modified	0.574	0.584	Pseudo-second order
Unmodified	0.765	0.770	Pseudo-second order
Copper			
Modified	0.512	0.502	Pseudo-first order
Unmodified	0.583	0.573	Pseudo-first order

Table 2. The Lagergren first-order and Ho et al. second-order data.

For all metals the adsorption rates are very fast initially and become almost constant as time increases showing that an equilibrium has been reached.

For cadmium and copper, the adsorption followed pseudo-first-order kinetics showing that only one molecule was involved in the rate determining step [43]. The adsorption of lead followed pseudo-second-order kinetics for both the ethylenediamine modified and the unmodified algal material. Similar results were obtained by Mwangi and Ngila [29] when studying the removal of heavy metals from wastewater using ethylenediamine-modified seaweed.

3.8. Analysis of wastewater samples

Fifty millilitre samples of water from Turi River were placed into plastic bottles and spiked with known concentrations of lead, cadmium and copper. The solution was agitated in a shaker for 30 min and then filtered, and the filtrate was analyzed for lead, cadmium and copper samples using AAS. The percentage of the metal recovered from the water samples was then recorded. Results from this analysis are shown in **Table 3**.

Metal	Concentration added/μgl^{-1}	Concentration recovered/μgl^{-1}	% Recovery
Copper			
Unmodified	00.00	2.050	
	10.00	10.363	86.0
	20.00	18.191	82.5
	30.00	19.871	62
Modified	00.00	2.010	
	10.00	8.191	68.2
	20.00	6.163	28.0
	30.00	7.682	24.0
Cadmium			
Unmodified	00.00	1.137	
	10.00	8.019	72.0
	20.00	14.627	69.2
	30.00	18.370	59.0
Modified	00.00	1.145	
	10.00	8.916	80.0
	20.00	16.070	76.0
	30.00	22.892	73.5
Lead			
Unmodified	00.00	2.185	
	10.00	9.870	81.0
	20.00	17.659	79.6
	30.00	19.633	61.0
Modified	00.00	2.070	
	10.00	9.885	81.9
	20.00	16.729	75.8
	30.00	22.449	70.0

Table 3. The percentage recovery of metals from real water samples.

From the results, the percentage recovery is high at low concentrations but decreased as the concentration of the metal in the water sample increased. The adsorption of copper by the unmodified sample was found to be the best and cadmium the least. For the modified sample, the adsorption of lead was the most and cadmium was the least. This can be explained by the fact that copper ions unlike lead and cadmium have relatively high affinity for ligands containing the nitrogen atom [44]. Such ligands which are smaller than the adsorbent have high affinity for the metal ion due to their high basicity [45]. These results show that the algal biomass has shown good potential to be used in water resource management.

4. Conclusions

The study successfully functionalized the algal material with tetramethylethylenediamine, and the FTIR spectrum provided evidence of its functional groups capable of binding metal ions. Adsorption of the three metals was best at lower pH values (4.2–6.8). Beyond these values, the adsorption decreased considerably. The rate of adsorption was very fast as more that 90% of the metals were removed from the solution within 30 min. The adsorption of copper fitted into the Langmuir adsorption isotherm with R^2 values of 0.974 and 0.512 for the unmodified and modified sorbents, respectively, indicating a monolayer-binding mechanism. Cadmium and lead fitted into the Freundlich adsorption mechanism (R^2 values were Cd modified = 0.993, unmodified = 0.251, Pb modified = 0.906, unmodified = 0.929). The adsorption of lead and cadmium was by Ho's pseudo-second-order kinetics confirming a multi-site interaction, whereas copper followed the pseudo-first-order kinetics, evidence of single-site adsorption. The adsorption by the algal material did not improve upon modification since natural ligands were replaced with ethylenediamine which has a lower stability constant for the metal analytes but minimized leaching of dissolved organic matter. In general the study has shown that the algal material can be used for as an effective sorbent for removal of lead, cadmium and copper from contaminated wastewater at low pH values.

Author details

John Okapes Joseph[1], Isaac W. Mwangi[1*], Sauda Swaleh[1], Ruth N. Wanjau[1], Manohar Ram[1] and Jane Catherine Ngila[2]

*Address all correspondence to: isaacwaweru2000@yahoo.co.uk

1 Department of Chemistry, Kenyatta University, Nairobi, Kenya

2 Department of Chemical Technology, University of Johannesburg, South Africa

References

[1] Love J., Percival E. (1964) The polysaccharides of the green seaweed codium fragile. The water-soluble sulphated polysaccharides. *Journal of Chemical Society* Part II 3338–3345

[2] Matsubara K., Matsuura Y., Bacic A, Liao M.L., Hori K., Miyazawa K. (2001) Anticoagulant properties of a sulfated galactan preparation from a marine green alga, *Codium cylindricum*. *International Journal of Biological Macromolecules* 28:395–399

[3] Selivanovskaya S.Y., Latypova, S., Latypova, V.Z. (2003) The use of bioassays for evaluating the toxicity of sewage sludge and sewage sludge-amended soil. *Journals of Soils and Sediments* 3 (2): 85–92.

[4] Henry, C.L., Cole, D.W. (1997). Use of bio-solids in the forest: technology, economics and regulations. *Biomass and Bioenergy* 13: 69–277

[5] Losada, M.R., Lopez-Diaz, L., Rodriguez, A. 2001. Sewage sludge fertilisation of a silvopastoral system with pines in northwestern Spain. *Agroforestry Systems* 53: 1–10.

[6] Gupta, V.K., Gupta, M., Sharma, S. (2001) Process development for the removal of lead and chromium from aqueous solution using red mud, an aluminum industry waste. *Water Research* 35: 1125–1134.

[7] Lodeiro, P., Barriada, J.L., Herrero, R., Sastre De Vicente, M.E. (2006) The marine macroalga Cystoseira baccata as biosorbent for cadmium (II) and lead (II) removal: kinetic and equilibrium studies. *Environmental Pollution* 142 (2): 264–273.

[8] Cha, D.K., Song, J.S., Sarr, D. (1997) Treatment technologies. *Water Environment Research* 69: 676–689

[9] Hsien, T.Y., Rorrer, G.L. (1995) Effects of acylation and crosslinking on the material properties and cadmium ion adsorption capacity of porous chitosan beads. *Separation Science and Technology* 30: 2455–2475.

[10] Chen, J.P., Yang, L. (2005) Chemical modification of Sargassum sp. for prevention of organic leaching and enhancement of uptake during metal biosorption. *Industrial and Engineering Chemistry Research* 44: 9931–9942.

[11] Saracoglu, S., Soylak, M., Elci, L. (2001) Preconcentration of Cu (II), Fe (III), Ni(II), Co(II) and Pb(II) ions in some manganese salts with solid phase extraction method using chromosorb-102 331resin. *Trace Elements and Electrolyte* 18:129–133.

[12] Marshall, W.E., Wartelle, L.H., Boler, D.E., Johns, M.M. and Toles, C.A. (1999) Enhanced metal adsorption by soybean hulls modified with citric acid. *Journal of Bio Resources Technology* 69:263–268.

[13] Guo, Y. J., Qi, S., Yang, K., Yu, Z., Wang, H. (2002) Adsorption of Cr (VI) on micro-and mesoporous rice husk-based activated carbon. *Journal for Materials for Chemistry and Physics* 78:132–137

[14] Bailey, S.E., Olin, T.J., Bricka, R.M., Adrian, D.D. (1999) A review of potentially low-cost sorbents for heavy metals. *Water Resources Journal* 33:2469–2479.

[15] Karsten, E., Erhard, H., Roland, R., Hartmut, H. (2005) Amines, Aliphatic in Ullmann's Encyclopedia of Industrial Chemistry, Wiley-VCH Verlag, Weinheim.

[16] Vogel, A.I. (1978) A Textbook of quantitative Inorganic Analysis Including Elementary Instrumental Analysis. 3rd ed. New impressions, London, p. 910, 1162

[17] Lagergren, S. (1898) The theory of so-called adsorption of soluble substances. Handlingar 24 (4): 1–39.

[18] Ho, Y.S., McKay, G., Wase, D.J., Forster, C.F. (2000) Study of the sorption of divalent metal ions on peat. *Adsorption Science Technology* 18: 639–650

[19] Langmuir, I. (1918) The adsorption of gases on plane surfaces of glass, mica and platinum. *Journal of American Chemical Society* 40: 1362–1403.

[20] Tan, C., Xiao, D. (2009) Adsorption of cadmium ion from aqueous solution by ground wheat stem. *Journal of Hazardous Materials* 164: 1329–1363.

[21] Yang, R.T. (1987) Gas separation by adsorption processes. *Journal of the American Chemical Society* 53: 497.

[22] Ng, J.C.Y., Cheung W.H., McKay, G. (2003) Equilibrium studies for the sorption of lead from effluents using chitosan. *Chemosphere* 52 (6): 1021–1030.

[23] Freitas, P.A., Iha, M.K., Felinto, M.C.F.C., Suárez-Iha, M.E.V. (2008) Adsorption of di-2-pyridyl ketone salicyloylhydrazone on Amberlite XAD-2 and XAD-7 resins:characteristics and isotherms. *Journal of Colloid and Interface Science* 323: 1–5.

[24] Freundlich, H.M.F. (1906) Over the adsorption in solution. *Zeitschrift für Physikalische Chemie* 57 (A) 385–470.

[25] Stuart, B. (1996) Modern Infrared Spectroscopy, John Wiley & Sons Inc., New York

[26] Matheickal, J.T., Yu, Q. (1996) Biosorption of lead from aqueous solutions by marine alga, Ecklonia radiate. *Water Science Technology* 34: 1–7.

[27] Kapoor, A., Viraraghavan, T. (1997). Heavy metal biosorption sites in *Aspergillus niger*. *Bioresources and Technology* 61: 221–227.

[28] Schluter, A. (1988) Poly ([1,1,1]propellane). A novel rigid-rod polymer obtained by ring opening polymerization breaking a carbon-carbon A-bond. *Journal of the American Chemical Society* 21(5): 1208–1211.

[29] Mwangi, I.W., Ngila, J.C. (2012) Removal of heavy metals from contaminated water using ethylenediamine-modified green seaweed, *Caulerpa serrulata*. *Physics and Chemistry of the Earth* 50: 111–120.

[30] Singh, S.P., Ma, L.Q., Hendry, M.J.. 2006 Characterization of aqueous lead removal by phosphatic clay: equilibrium and kinetic studies. *Journal of Hazardous Materials* 136 (3): 654–662

[31] Benhima, H., Chiban, M., Sinan, F., Seta, P., Persin, M. (2008) Removal of lead and cadmium ions from aqueous solution by adsorption onto micro-particles of dry plants. *Colloids Surface Biochemistry: Biointerfaces* 61: 10–16.

[32] Godhane, I., Nouri, L., Hamdaoui, Q., Chiha, M. (2008) Kinetic and equilibrium study for the sorption of cadmium (II) ions from aqueous phase by eucalyptus bark. *Journal of Hazardous Materials* 152: 148

[33] Richardson, J.T. (1967) Crystal field effects in ion-exchanged faujasites. *Journal of Catalysis* 9(2): 178–181.

[34] Tripath, T., Ranjan-De, B. (2006) Flocculation: A New Way to Treat the Waste Water, *Journal of Physical Sciences*, 10:93 –127

[35] Li, X., Tang, Y., Cao, X., Lu, D., Luo, F., Shao, W. (2008) Preparation and evaluation of orange peel cellulose adsorbents for effective removal of cadmium, zinc, cobalt and nickel, colloids and surfaces. *Physicochemical and Engineering Aspects* 317: 512–521.

[36] Keskinkan, O., Goksu, M.Z.L., Basibu Yuk, M., Forster, C.F. (2004) Heavy metal adsorption properties of a submerged aquatic plant *Ceratophyllum demersum*. *Bioresource Technology* 92: 197

[37] Yang, J.B., Volesky, B. (1999) Cadmium biosorption rate in protonated Sargassum biomass. *Environmental Science and Technology* 33: 751.

[38] Khani, M. H. (2006) Biosorption of uranium from aqueous solutions by nonliving biomass of marine algae, *Cystoseira indica*. *Journal of Biotechnology* 9(2): 101–108.

[39] Ilhan, S.N.M., Kilicarslan, S., Ozdag, H. (2004) Removal of chromium, lead and copper from industrial waste by *Staphylococcus saprophyticus*. *Turkish Electronic Journal of Biotechnology* 2: 50–57.

[40] Puranik, P.R., Paknikar, K.M. (1997) Biosorption of lead and zinc from solutions using *Streptoverticillium cinnamoneum* waste biomass. *Journal of Biotechnology* 55: 113.

[41] Bradl, H.B. (2004) Adsorption of heavy metal ions on soils and soils constituents. *Journalof Colloid and Interface Science* 277: 1–18

[42] Drake, L.R., Lin, S., Rayson, G., Jackson, P.J. (1996) Chemical modification and metal binding studies of *Datura innoxia*. *Environmental Science and Technology* 30: 110–114.

[43] Agrawal, A., Sahu, K.K. (2006) Kinetics and isotherm studies of cadmium adsorption on manganese nodule residue. *Journal of Hazardous Materials* 137(2): 915–924.

[44] Topperwien, S.B.R., Xue, H., Asigg, A. (2007) Cadmium accumulation in *Scenedesmus vacuolatus* under fresh water conditions. *Journal of Environmental Science and Technology* 41:5383–5388.

[45] Comuzzi, C., Grespan, M., Melchior, A., Portanova, R., Tolazzi, M. (2001) Thermodynamics of complexation of cadmium (II) by open chain N-donor ligands in dimethyl sulfoxide solution. *European Journal of Inorganic Chemistry* 12: 3087–3094.

3

Integrated Approaches in Water Quality Monitoring for River Health Assessment: Scenario of Malaysian River

Salmiati, Nor Zaiha Arman and Mohd Razman Salim

Additional information is available at the end of the chapter

Abstract

Current practice of determining river water quality in Malaysia is based mainly on physicochemical components. Perhaps, owing to the lack of information on habitat requirements and ecological diversity of aquatic macroinvertebrates and on unearthly taxonomic key of benthic macroinvertebrates in this region makes it less popular than conventional methods. The study took place in three rivers in the state of Johor, Southern Peninsula of Malaysia, which exhibited different degrees of disturbances and physical properties, namely Sungai Ayer Hitam Besar, Sg Berasau, and Sg Mengkibol. Benthic macroinvertebrates were sampled using rectangular dipnet with frame dimension 0.5 m × 0.3 m. Although physicochemical elements such as water temperature, pH, and dissolved oxygen (DO) were measured using a YSI Professional Plus handheld multiparameter instrument, other parameters such as biochemical oxygen demand (BOD_5), chemical oxygen demand (COD), total suspended solid (TSS), and ammoniacal nitrogen (NH_3N) were tested using the procedure of APHA Standard Method. The study found that the status of water quality varies among the three rivers. A multivariate analysis, the canonical correspondence analysis (CCA), was applied to elucidate the relationships between biological assemblages of species and their environment using PAST (version 2) software. The present findings reveal that human-induced activities are the ultimate causes of the alteration in macroinvertebrate biodiversity.

Keywords: benthic macroinvertebrate, ecosystem health, anthropogenic impact, water quality monitoring, biological assessment

1. Introduction

Water is a natural resource that is vital to all life-forms. Although nearly 70% of the world is covered by water, only 2.5% of the total is freshwater. The rest is ocean-based saline water. However, only 1% of the freshwater is easily accessible, with much of it trapped in glaciers and snowfields. In tandem with the growing global population and improvement of living standards, the increasing demand for freshwater has been said to overshadow the concerns of the warming effect of climate change [1].

Since time immemorial, rivers have played a major role in the development of human society, serving as transport routes and as a vital supply of water for domestic and agricultural use, while yielding an important source of protein for human consumption. Hence, it is not surprising that many major towns and cities are situated on the banks of rivers. For example, early urban settlements such as Uruk, Eridu, and Ur, established at the dawn of human civilization about 6000 years ago (4000 BC) in Mesopotamia and Babylon, were built in the fertile valley irrigated by the Tigris and Euphrates rivers [2].

2. Scenario of river in Malaysia

Rivers have similarly played an important role in the growth of towns and cities in Malaysia, with early settlements springing up along river banks and estuaries [3]. Many major cities and towns in such locations include Kuala Lumpur, Kuala Terengganu, Alor Setar, Kuantan, Kota Bharu, Kuching, and Melaka City [4, 5]. The discovery of tin deposits in the flood plains and river valleys also encouraged settlements to mushroom in these areas, leading to a booming tin-mining industry in the 1800s till 1980s, which made the country the largest producer of tin in the world.

Malaysia has grown rapidly over the last three decades, transforming from a rural economy based on agriculture and tin mining to an export-based, manufacturing economy. In the eighteenth century and the first half of the nineteenth century, large areas of land were cleared for coffee and sugarcane cultivation. This was followed by large-scale land clearing for rubber plantations, making Malaya the world's largest producer of natural rubber. In recent years, much of the rubber growing lands has been converted to oil palm cultivation, while further new areas have been cleared for this crop. Unfortunately, rapid changes of land use, especially of forested land and food crops to plantations as well as urban development, have triggered river erosion, surface runoff, and sedimentation of rivers, resulting eventually in overstressed river systems. River basins are frequently facing problems arising from flooding. Many rivers are gradually losing their ability to supply fresh water, and as a result, these rivers are now mainly used for transportation [6].

In Malaysia, the sources of raw fresh water are rivers, storage dams, and groundwater. Rivers supply 90% of the nation's water supply, providing water for various uses such as domestic, agricultural and industrial processes, power generation, besides serving as waterways for transport and communication. Aquatic harvests from rivers are also an important source of food. However, as the country develops, water pollution is becoming more serious, affecting the function of the river system as a source of raw water supply. Although raw water supply is not yet depleted, clean water that can be safely consumed by humans is becoming hard to come by.

The major causes of water pollution in Malaysia include effluent from wastewater treatment plants, discharge from agro-based industries and livestock farming, land clearing activities, and domestic sewage [7]. Rivers in both urban and rural areas are experiencing the same problems. Although environmental issues in Malaysia raise serious concerns, the measures taken to address the problem thus far have been fragmented and inadequate. An integrated and holistic approach that is required is now gaining recognition, and this is reflected in the government's latest policies.

3. River-related issues and river management in Malaysia

As a responsible authority for ensuring the sustainability of integrated river basin and water resources management, the Department of Irrigation and Drainage (DID) under the Ministry of Natural Resources and Environment (NRE) upheld the Integrated River Basin Management (IRBM) concept more than 10 years ago. IRBM, a subset of Integrated Water Resources Management (IWRM), is an effective method or approach to achieve the objectives of the IWRM-based river basin. In other words, IRBM is the management of river basin as an entity, not as a series of isolated individual rivers. It is geared towards integrating and coordinating policies, programs, and practices in addressing water and water-related issues. It also requires the improvement of professional and financial practices as well as legislative, managerial, and political capacity on water-related issues. The One State One River Program (1N1S), launched by DID in 2005, was an extension of the Love Our Rivers Campaign with the slogan "Sungaiku Hidupku" ("My river, My life"). This program is one of the pilot projects for the implementation of the IRBM concept. In this program, DID and the state governments selected 13 rivers, one river for each state. Among the main criteria of the river selection was that the polluted rivers should be running through major cities in the country. The main goals of the 1N1S were to achieve and maintain the status of clean and vibrant river within Class IIB of water quality by 2015 [8]. Under the RMK-9, a sum of RM57.5 million was allocated to each state, while in the RMK-10, an allocation of RM26 million was provided for a period of 2 years (2011–2012) for 13 selected rivers.

The results showed that the program had achieved some measure of success, especially in terms of improved water quality from Class V to Class III in some rivers, namely Sungai Petani, Kedah; Sungai Galing, Pahang and Sungai Pinang, Pulau Pinang. In addition, Sungai Kinta in Perak achieved an improvement in water quality index (WQI) from Class III to Class IIB. However, the water quality for Sungai Hiliran, Terengganu and Sungai Penchala, Kuala Lumpur remain unchanged [8]. Those river restoration programs have not only shown positive effects and significantly improved the quality of water, but also enhanced amenities and riverside landscape. Nevertheless, the positive effects of the measures on riverine biota are rarely observed or documented. Sungai Melaka, for example, has shown tremendous changes in water quality, from heavily polluted to slightly polluted after undergoing several rehabilitation efforts. However, in terms of faunal diversity and aquatic life, only tolerant and hardy species such as the tilapia fish have been found to inhabit the river. A similar situation also occurs in two rehabilitated rivers in Johor, namely Sungai Sengkuang and Sungai Sebulong, where only hardy, non-economic fish species have been observed.

However, taking into consideration the physicochemical aspects alone are not sufficient to indicate a healthy ecosystem as a whole. In fact, this does not guarantee health of aquatic life because it does not directly reflect the biological responses to pollution. Although physico-chemical evaluation might be appropriate to particular circumstances at the time of sampling, it does not provide an insight into the effects of pollution on habitat and aquatic life. Aquatic communities respond to ecosystem changes in various ways. The distribution and abundance of certain species and changes in their behavioral, physiological, and morphological of individual organisms indicate whether that habitat has been adversely altered. High biodiversity of aquatic species and the presence of sensitive species are good signs of a healthy stream. Nature of the river as a collection point for water flowing from every corner reflects the health of the surrounding area. Therefore, any changes or modification on riparian vegetation and surrounding landscape may subsequently alter the composition and functional structure of aquatic life inhabiting it. Healthy water body shows ecological integrity, which represents the natural or undisturbed area. Ecological integrity is a combination of three components, namely chemical, physical, and biological integrities. When one or more of these components are degraded, the health of the water body is affected, and in most cases, aquatic life living in it will reflect the degradation. According to Gordon et al. [9], stream health measurement takes into consideration the water quality, habitat availability and suitability, energy sources, hydrology, and the biota themselves.

In order to achieve a comprehensive evaluation of healthy water bodies, biological assessment tool should be carried out simultaneously with the standard physicochemical method. Biological assessment, the primary tool to evaluate the biological condition of a water body, comprises surveys and other direct measurements through biological communities such as plankton, periphyton, microphytobenthos, macrozoobenthos, aquatic macrophytes, and fish. Among all, benthic macroinvertebrates are the most favored in freshwater monitoring and are widely used to evaluate the water body health and condition [10, 11]. The advantages of using biological indicators, particularly macroinvertebrates, are biological communities that reflect the overall ecological quality and provide a broad measurement of fluctuating environmental conditions. In addition, the result of biological monitoring is reliable and relatively inexpensive compared to toxicity testing [12]. Liebmann (1962) quoted that the history of biological monitoring methods for assessing water quality began more than a century ago by Kolenati (1848) and Cohn (1853) both quoted by [13]. However, such studies in Malaysia are still very limited and started relatively late with the earliest documented was in the early 90s [14, 15]. After year 2000, interest on this topic is gaining attention and grows, and example of studies can be seen in [16–20]. In the year 2009, DID in collaboration with Universiti Sains Malaysia produced a Guideline for Using Macroinvertebrates for Estimation of Streams Water Quality. The guideline provides simple, inexpensive, and easy approach to estimate water quality through the identification of freshwater macroinvertebrates. This government's effort is an initial step to the development of such studies in Malaysia and proving biological methods in the study of water quality began to be accepted.

4. Description of study area

This study focused on the description of the existing ecological environment of three rivers with different environmental gradient ecosystem, *viz.* Sg Ayer Hitam Besar (forest reserve), Sg Berasau (logged area), and Sungai Mengkibol (urban and rehabilitated rivers). Three main processes explored in this study were consisted of physical characteristics (general characteristics that are important in influencing the river's aquatic ecology such as channel forms, instream habitats, substrates, riverbank vegetation, and structure; additional habitat attributes such as anthropogenic alterations to the river were briefly described), biological characteristics (focusing on the composition and abundance of macroinvertebrates species) and chemical characteristics (documentation of existing conditions related to commonly observed water quality parameters). The study also investigated the correlation between the physicochemical attributes and variations in the macroinvertebrates assemblages.

4.1. Sg Ayer Hitam Besar

Sg Pontian Besar drains a total area of 362.047 km² to the Straits of Malacca. The main tributaries are Sg Ayer Hitam Besar and Sg Rambutan. The Sg Ayer Hitam Besar subcatchment is situated between Ulu Pontian and Kampung Seri Gunung Pulai settlement area. This river has a length of 11.2 km with a width of 35 m, before joining the main river and end up at the Straits of Malacca. The study was conducted at the Pulai Waterfall in Gunung Pulai Forest Reserve, in the southwestern part of Johor. Located within an 8-ha protected forest reserve, this former water catchment area for Singapore serves an important function as an area for habitat and biodiversity protection, recreation, tourism, research, and education. Gunung Pulai is a hill dipterocarp forest type on granite-based soil; its peak is about 700 m above sea level. The Gunung Pulai Forest Reserve is divided into 27 compartments to facilitate administration and management. However, only two compartments, namely Compartment 9 for recreational pursuits and Compartment 7 for educational purposes, are open to the public.

4.2. Sg Berasau

Sg Berasau is located in the Kota Tinggi District, about 42 km northeast of Johor Bahru. With an area of 54 km², this river basin is a sub-basin of Sungai Ulu Sedili Besar river basin. The study area was a first-order river situated about 8 km from the Kota Tinggi-Jemaluang main road and within the Ulu Sedili Forest Reserve. The sampling areas could be reached only by 4WD through forest plantations managed by Aramijaya Sdn. Bhd., which granted the study team access to the area for this study. Sg Berasau is surrounded by a permanent forest reserve.

However, there is active logging in the area planted with *Acacia mangium,* a tree species of the pea family, Fabaceae. *Acacia mangium* is suitable as raw material for sawn timber, and wood-chips in the pulp and paper industry, and reconstituted wood for the furniture industry. In the Asia–Pacific region, Japan is among the larger importers of wood chips, while the pulp and paper industry finds markets in Taiwan and South Korea. Based on the macro-EIA for forest management units (FMU) in Johor, Sg Berasau supports a variety of fish species such as Terbul, Sebarau, Baung, Seluang, and Tapah.

4.3. Sg Mengkibol

Sg Mengkibol is a second-order river located within the Endau watershed. This river receives flows from Sg Melantai before joining Sg Semberong. This river basin is approximately 185 km^2 width and 20 km long. The study area is located in the middle section of Sg Mengkibol, starting from the Sg Mengkibol Riverine Park until the wet market. From December 2006 until January 2007, Malaysia experienced large-scale flood events that affected most of the state of Johor. This phenomenon was caused by extremely high rainfall attributed to Typhoon Utor that made landfall in the Philippines and Vietnam. A series of massive floods hit the states of Malacca, Pahang, and Negeri Sembilan, with Johor as the worst hit state. Among the major towns affected by the floods were Batu Pahat, Johor Bahru, Kluang, Kota Tinggi, Mersing, Muar, Pontian, and Segamat. Consequently, the Department of Irrigation and Drainage (JPS) allocated RM2 million to deepen Sg Mengkibol at its banks for flood mitigation in low-lying residential areas [21].

5. Sampling procedure and data collection

Most of that data used for analyses in the study were primary data obtained from sampling and laboratory analyses. Data of benthic macroinvertebrate assemblage and water quality assessment were obtained from *in situ* analyses and further analyses in the laboratory. River habitat and morphology data were measured on-site in field surveys. On the other hand, secondary data such as land use, rainfall data, and stream catchment maps were obtained from several agencies such as the Department of Environment, Department of Drainage and Irrigation, Department of Survey and Mapping and local authorities.

5.1. Benthic macroinvertebrate

The multihabitat approach of USEPA's Rapid Bioassessment Protocol (RBP) was adopted for this study as it was suitable for sampling a wide variety of stream types [22]. In this study, a rectangular dipnet with 500-µm mesh attached to a 0.5 m × 0.3 m frame and a long pole were used for this purpose. For multiple habitats, the habitat types were sampled in proportion to their relative surface area within the sampling reach. A total of 20 sample units were collected from all major habitat types by kicking the substrates or jabbing with a dipnet within a sampling station to obtain a composite of 60 sample units in total. The samples were washed and any detritus present was removed on-site as it would be impractical to wash large samples in the laboratory. Following this, benthic materials were sieved and rinsed before preservation in 70% ethanol. The sample containers were labeled to show all the essential information, including date and sampling location. Preprinted labels were preferably used using marker pens as ethanol would remove writing. Moreover, labeling on container lids was avoided in case they were interchanged. In laboratory, benthic macroinvertebrates were rinsed thoroughly in 500 μm-mesh sieves to remove preservative and sediment, while remaining debris was visually inspected and discarded. During the separation process, the samples were soaked for about 15 min in tap water to hydrate the preserved organisms and prevent them from floating on the water surface during sorting. The samples were then spread over an enamel tray and sorted out into major taxa. All organisms were identified to the lowest practical level using a dissecting microscope and taxonomic Key from Yule and Yong [23].

5.2. Water quality

Water quality assessment data were obtained by two methods, namely *in situ* and laboratory analyses. *In situ* measurements were made on temperature, pH, conductivity, and dissolved oxygen (DO) by using a YSI Professional Plus handheld multiparameter instrument. Meanwhile, other parameters such as biochemical oxygen demand (BOD_5), total suspended solid (TSS), ammoniacal nitrogen (NH_3N), and chemical oxygen demand (COD) were measured in the laboratory based on the Standard Methods for the Examination of Water and Wastewater [24]. Both *in situ* readings and water samples were collected from the same location. Water samples were collected in 1-L polyethylene bottles and chilled in a cold box filled with ice cubes (4°C) to minimize the metabolism of organisms contained in the water. The water samples were labeled in a manner similar to that used for the benthic samples.

The water quality index (WQI) was calculated to indicate the level of pollution and the corresponding suitability for use according to the National Water Quality Standards for Malaysia (NWQS). Water quality class was determined based on the water quality index (WQI), ascertained by the six parameters, *viz.* pH, DO, BOD_5, COD, TSS, and NH_3N, according to the DOE formula (1):

$$WQI = (0.22*SIDO) + (0.19*SIBOD) + (0.16*SICOD) + (0.15*SIAN)$$
$$+ (0.16*SISS) + (0.12*SIpH) \tag{1}$$

where SI is the subindex of the respective water quality parameters which is used to calculate the WQI (**Table 1**). The WQI classification based on water use is shown in **Table 2**.

Subindex for DO (in % saturation)	
SIDO = 0	for x ≤ 8
SIDO = 100	for x ≥ 92
SIDO = −0.395 + 0.030x^2 − 0.00020x^3	for 8 < x < 92
Subindex for BOD	
SIBOD = 100.4 − 4.23x	for x ≤ 5

Subindex for DO (in % saturation)	
SIBOD = 108 × exp(−0.055x) − 0.1x	for x > 5
Subindex for COD	
SICOD = −1.33x + 99.1	for x ≤ 20
SICOD = 103 × exp(−0.0157x) − 0.04x	for x > 20
Subindex for NH3-N	
SIAN = 100.5 − 105x	for x ≤ 0.3
SIAN = 94 × exp(−0.573x) − 5 × |x − 2|	for 0.3 < x < 4
SIAN = 0	for x ≥ 4
Subindex for SS	
SISS = 97.5 × exp(−0.00676x) + 0.05x	for x ≤ 100
SISS = 71 × exp(−0.0061x) − 0.015x	for 100 < x < 1000
SISS = 0	for x ≥ 1000
Subindex for pH	
SIpH = 17.2 − 17.2x + 5.02 × 2	for x < 5.5
SIpH = −242 + 95.5x − 6.67 × 2	for 5.5 ≤ x < 7
SIpH = −181 + 82.4x − 6.05 × 2	for 7 ≤ x < 8.75
SIpH = 536 − 77.0x + 2.76 × 2	for x ≥ 8.75

Table 1. Best fit equations for the estimation of various subindex values.

Class	Uses
Class I	Conservation of natural environment
	Water supply I—practically no treatment necessary
	Fishery I—very sensitive aquatic species
Class IIA	Water supply II—conventional treatment required
	Fishery II—sensitive aquatic species
Class IIB	Recreational use with body contact
Class III	Water supply III—extensive treatment required
	Fishery III—common, of economic value and tolerant species; livestock drinking
Class IV	Irrigation
Class V	None of the above

Table 2. Water classes and uses.

5.3. Characterization of river habitat

A visual-based habitat assessment based on the USEPA habitat assessment survey was carried out simultaneously with the biological sampling. Several features in habitat assessment include a general description of the site, a physical characterization and water quality assessment, and also a visual assessment of instream and riparian habitat quality. Physical characterization comprised documentation of general land use, description of the stream origin and type, and a summary of the riparian vegetation features that included measurements of instream parameters such as width, depth, flow, and substrates. The observed channel dimensions were carried out in the survey stretch, located either in the presence of riffle or at a suitable shallow section of the river. Channel dimension measurements were taken according to the river habitat survey (RHS) method.

5.4. Streamflow gauging

Streamflow gauging was conducted to measure the flow rate of the study area. The equipments used in this study were flow meter, measuring tape, staff level, hammer, and ropes/cables. The Cole Parmer Model BS 11000 flow meter used in this study was equipped with a propeller to allow it to rotate according to the velocity of the water. The mean section method was used to measure the river discharge using the flow meter and other tools, whereby cross-sectional area of the river was divided into several subsections. Stream velocity was measured at depths of 0.6d, 0.2d, and 0.8d depending on the need, where d represents the variable of depth from the water surface. This method was used to obtain the average velocity of the represented river. The general hydraulic formula for river discharge is as follows (2)

$$Q = AV \tag{2}$$

where Q is discharge (volume/unit time-e.g. m^3/s, also called cumecs), A is the cross-sectional area of the stream (e.g. m^2), and V is the average velocity (e.g. m/s).

6. Results and discussion

6.1. Benthic macroinvertebrate compositions

Overall, a total of 1081 fauna were recorded from Sg Ayer Hitam Besar. On the other hand, 610 individuals were caught from Sg Berasau, while sampling from Sg Mengkibol documented 1008 individuals. This brings the total number of macroinvertebrates assemblage based from all sampling events of 2699 individuals. Common netspinners larvae from caddisfly family dominate the research finding in Sg Ayer Hitam Besar, contributing approximately one-third of the total samples (30.52%). Some intolerant taxa represented by Plecopteran families were seen widely distributed in the study area. They are Perlidae (242 individual) and Chloroperlidae (194 individual). Interestingly, Decapods (Palaemonidae) has been found widely distributed with relatively high abundance and dominated in every sampling session. On the other hand, mollusks were found in very low composition restricted to Physidae and Pleuroceridae (**Table 3**). Trichoptera represents the percentage of the most abundant and adaptive species with the most dominant family of Hydropsychidae. This higher number is believed to be associated with the presence of algal biomass [25]. This insect tends to live in sheath made from organic debris and mineral fragments and makes the surface of the substrate as their habitat. This insect larvae are also often attached to rocks, facing the flow and feed on the particles trapped in their nets [26]. Several groups of aquatic insects favored rocky substrate as it offers habitat for protection and oviposition [27]. In this present study, sufficient numbers of oviposition sites were observed, including plenty of rocky substrates and riverbank vegetations, which could explain the high abundance of caddisfly larvae in this area.

Order	Family	Abundance	Percentage (%)
Decapoda	Palaemonidae	153	14.15
	Potamidae	5	0.46
Ephemeroptera	Heptageniidae	57	5.27
	Ephemeridae	3	0.28
Plecoptera	Perlidae	242	22.39
	Capniidae	3	0.28
	Chloroperlidae	194	17.95
Trichoptera	Hydropsychidae	305	28.21
	Limnephilidae	12	1.11
	Polycentropodidae	6	0.55
	Leptoceridae	7	0.65
Coleoptera	Elmidae	60	5.55
	Pyralidae	6	0.56
Odonata	Calopterygidae	1	0.09
	Lestidae	2	0.19
	Gomphidae	3	0.28
	Libellulidae	16	1.48
Gastropoda	Pleuroceridae	1	0.09
	Physidae	4	0.37
Hemiptera	Veliidae	1	0.09
	Total	**1,081**	**100**

Table 3. Benthic macroinvertebrate compositions in Sg Ayer Hitam Besar.

Decapoda exhibited the highest distribution with an abundance of 343 individual in Sg Berasau. Both families, Palaemonidae and Potamidae, contributed more than half (56.23%) from the total amount. Caridean prawn (Palaemonidae) were found in all sampling events with large numbers compared to others (291 individual). The second group which had the highest distribution was Odonata, with an abundance of 153 individual. Gomphidae, which belongs to Anisoptera suborders, were the second largest taxa found, consisting of 91 individuals. The least dominant families in Sg Berasau were Capniidae, Sialidae, and Pyralidae, contributing 0.16% each from total percentage (**Table 4**). Freshwater prawns of the genus *Macrobrachium* are free-living decapod crustaceans, present in almost all permanent water bodies. They inhabit a wide variety of habitat even in extreme condition, where waters can reach pH 3.3, and stagnant pool with daytime temperature may reach 35°C [28]. Their feeding habits are variable, with some are scavengers or being detritivorous [29]. As such, they are very important in recycling organic matter in the environment. The inclusion of organic matter into water bodies due to logging activities in Sg Berasau is beneficial to shrimp, as can be seen from the abundance of these organisms. However, they are prone to human disturbance and development and become extinct. This happened in Sg Gombak, whereby the populations of *Atyopsis* species are now very rare due to rapid development, resulting in water pollution.

Order	Family	Abundance	Percentage (%)
Decapoda	Palaemonidae	291	47.70
	Potamidae	52	8.52
Ephemeroptera	Heptageniidae	30	4.92
	Ephemerellidae	18	2.95
	Baetidae	13	2.13
	Leptophlebiidae	2	0.33
	Potamanthidae	3	0.49
Plecoptera	Perlidae	11	1.80
	Capniidae	1	0.16
	Nemouridae	8	1.31
	Leuctridae	2	0.33
	Perlodidae	2	0.33
	Chloroperlidae	4	0.66
Odonata	Calopterygidae	17	2.79
	Lestidae	3	0.49
	Gomphidae	91	14.92
	Libellulidae	42	6.89
Gastropoda	Pleuroceridae	8	1.31
Hemiptera	Belostomatidae	2	0.33
	Nepidae	2	0.33
Megaloptera	Sialidae	1	0.16
Coleoptera	Hydrophilidae	6	0.98
	Pyralidae	1	0.16
	Total	**610**	**100**

Table 4. Benthic macroinvertebrate compositions in Sg Berasau.

Sampling of macrobenthic assemblages from Sg Mengkibol consists of moderately intolerant to very tolerant families. Odonates are on top of the list with highest abundance (448 individuals). Chironomidae or blood worm dominated the overall findings with cumulated percentage 22.32%. Gastropods, physidae, are in the second place with slight difference of cumulated percentage (21.92%). Odonates represented by Lestidae and Libellulidae also donated a relatively high number of 260 individuals. Interestingly, sensitive taxa were found in this study area, although the percentage is very low. They are mayflies and stoneflies (**Table 5**). Fly larvae can be found in various aquatic habitat and survived in most conditions. According to Yule [30], Chironomidae is probably the most diverse and abundant group of all stream macroinvertebrates. Chironomus, for example, were widely distributed in polluted areas [20, 31]. Hemoglobin pigment helps *Chironomus* spp. to adapt to unfavorable condition, since hemoglobin helps to sustain aerobic metabolism under low oxygen conditions [32]. Most fly larvae eat dead or dying plant and animal materials.

Order	Family	Abundance	Percentage (%)
Decapoda	Palaemonidae	20	1.98
Ephemeroptera	Ephemerellidae	1	0.10
	Baetidae	1	0.10
Plecoptera	Leuctridae	15	1.49
	Perlodidae	1	0.10
Coleoptera	Psephenidae	2	0.20
Odonata	Calopterygidae	60	5.95
	Lestidae	149	14.78
	Gomphidae	77	7.64
	Libellulidae	111	11.01
	Aeshnidae	41	4.07
	Coenagrionidae	10	0.99
Gastropoda	Pleuroceridae	3	0.30
	Physidae	221	21.92
	Viviparidae	4	0.40
Hemiptera	Naucoridae	35	3.47
	Nepidae	1	0.10
Hirudinea	Hirundinidae	28	2.78
Diptera	Chironomidae	225	22.32
	Syrphidae	3	0.30
	Total	**1,008**	**100**

Table 5. Benthic macroinvertebrate compositions in Sg Mengkibol.

6.2. River habitat survey

The Sg Ayer Hitam riverbed comprises more than 70% of natural structures such as cobble (riffles), large rocks, fallen trees, logs, and branches. These optimal conditions allow colonization, refugia, feeding/ spawning sites for aquatic faunal. Both Sg Ayer Hitam Besar and Sg Berasau consist of all four velocity/depth regime present in their study reach. The occurrence of slow-deep, slow-shallow, fast-deep, and fast-shallow velocity patterns reflects of habitat diversity and ability of stream to provide and maintain balance aquatic habitat. No channel alteration or dredging works present at studied reach in both rivers. Sg Ayer Hitam Besar showed and optimal condition of vegetative protection, as it covers more than 90% of streambank surface with native vegetation including trees, understory shrubs, or non-woody macrophytes. An optimal condition of vegetative zone serves as a buffer to pollution and nutrient input to the stream runoff, other than erosion control. Meanwhile, around 50–70% of the streambank surface is covered by riparian vegetation in Sg Berasau. Logging activities leave an obvious disruption as cropped vegetation/ bare soil potentially prone to high potential of streambank erosion during heavy downpour (30–60%).

Meanwhile, Sg Mengkibol exhibits an unsatisfactory habitat quality. Historically, Sg Mengkibol was hit by massive flood event in late 2006. In relation to deal with the incident over and over again, upgrading the river system for flood mitigation project has been carried out. Among the works are dredging and sediment disposals, as well as strengthening the river channel. As a result, variety of natural structure less than desirable due to frequent disturbed of epifaunal substrate. Compared to Sg Ayer Hitam Besar and Sg Berasau, shallow pools are more prevalent than deep pools at these rivers. Percentage of deposition of sediments in the Sg Mengkibol is approximately 50–80%, composing of gravel, sand, or fine sediment on the old and new bar. Increasing level of sediment deposition is an indication of instability and changing environment, thus unsuitable for many organisms. Deepening and dredging works as a part of river rehabilitation and restoration process have changed the shape of the stream channel drastically. More than 80% of stream reach has been straightened with the construction of anti-erosion measure in both sides of the banks. Straightened channel decreases the stream length 1–2 times shorter than its natural state. Channel sinuosity provides diverse habitat and fauna, as well as being able to handle surges as a result of storm. The construction of slope stabilization is carried out to reduce the amount of erosion that is likely to occur. Nevertheless, those artificial structures prevent plants from growing on streambanks. Therefore, the natural habitat for aquatic organism is limited. Riparian zone serves as a buffer to prevent the entrance of nutrients and pollutants directly into rivers. However, for urban river, riparian vegetative zone width is usually <6 m, due to extensive use of impervious surfaces. Therefore, it increases the volume of runoff and decreases groundwater recharge.

6.3. Water quality index (WQI)

A water quality index representing a gradation number describes the overall water quality in particular location and time based on several water quality parameters. The use of this index is not intended specifically for human health or aquatic life regulation, but provides simple guidance on water quality based on some important parameters. In Malaysia, the assessment and classification of water quality status are based on the water quality index (WQI) and the National Water Quality Standards (NWQS), which eventually grouped into certain classes. Index developed for Malaysia, the WQI is ascertained by six parameters, *viz.* pH, DO, BOD_5, COD, TSS and NH_3N. As summarized in **Figure 1**, overall mean WQI for Sg Ayer Hitam Besar was at Class I, indicating as an excellent quality. The mean values for each sampling event were ranged 90.67–97.00. Based on this index, Class I is defined as naturally very clean and preserved river. Its water resources are suitable as drinking water with minimal treatment. In terms of ecology, habitats are able to accommodate very sensitive aquatic species. Different situation is observed for Sg Berasau, although surrounded by a natural environment, land use activities such as deforestation have greatly affected the ecosystem health. Its effect can be seen through water quality status, which categorized this river into Class II (Clean). Based on general rating scale of WQI, Class II of water resources still can be used as a source of drinking after conventional treatment method. It is also suitable for recreational use with body contact. On average, the mean values of WQI in Sg Berasau are between 75.00 and 89.33 . Result of this study coincides with the finding from [17] which proves that logging activities which comply with prescribed standards still have an adverse impact on the riverine ecosystem even in a small proportion. On the other hand, Sg Mengkibol exhibits a moderately clean river status. Eight out of ten sampling events showed water quality for this river was in Class III (slightly polluted). This type of river status requires an extensive treatment as a drinking water supply. This river also accommodates certain fish species that are more tolerant and low in economic value such as catfish (*Clarias batrachus*) and tilapia (*Tilapia mossambica*) [33].

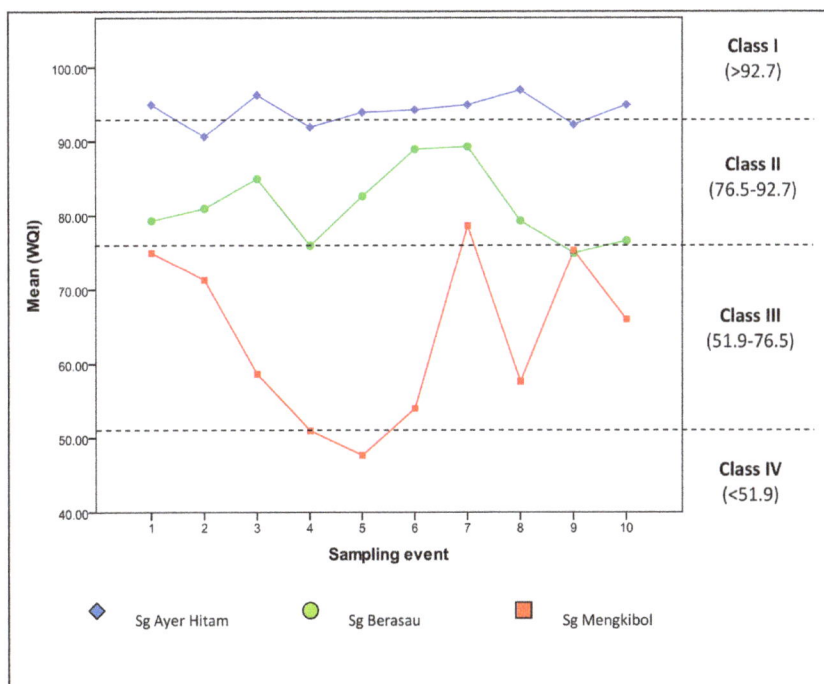

Figure 1. WQI values for all sampling sites.

6.4. Influence of hydrological, physicochemical, and habitat characteristics on the biological assemblage

In this section, a multivariate method, the canonical correspondence analysis (CCA), was applied to elucidate the relationships between biological assemblages of species and their environment using PAST (version 2) software. The CCA ordination biplot illustrated the relationship between several hydrological, physicochemical parameters, and distribution of the aquatic macroinvertebrates. The first two axes derived from CCA model accounted for 88.57% of the macroinvertebrates–environmental variations. CCA demonstrated that axis 1 was strongly correlated with DO, habitat quality (epifaunal substrate and vegetative protection), whereas COD, BOD$_5$, NH$_3$N, temperature, and velocity were negatively correlated with it (**Table 6**). Several taxa associated with three pollution-sensitive orders, EPT, showing good adaptation features and presenting highest score on the first axis. They consist of order Ephemeroptera (Heptageniidae, Ephemeridae), Plecoptera (Capniidae, Perlidae, Chloroperlidae), and Trichoptera (Hydropsychidae, Polycentropodidae, Leptoceridae, Limnephilidae). Low habitat quality, deterioration of DO and elevated concentration of nutrient and organic pollutant, suspended particulate, and temperature were positioned on the negative side of the first axis and were associated with moderate to tolerant taxa such as Odonata (Gomphidae, Libellulidae, Calopterygidae, Lestidae, Coenagrionidae, Aeshnidae), Hemiptera (Nepidae, Naucoridae), Coleoptera (Psephenidae), Gastropoda (Physidae, Planorbidae, Viviparidae), Diptera (Chironomidae, Syrphidae, Simuliidae, Tipulidae, Culicidae) and Hirundinidae (Hirundinidae). The second axis was positively related to TSS and pH. In particular, these species taxa (Pleuroceridae, Perlodidae, and Palaemonidae) showed moderate preference for velocity, water temperature, rainfall precipitation, and less vegetative cover. However, this group showed dependence on high suspended particulate and alkalinity.

Variable	Axis 1	Axis 2
Temperature	−0.891	−0.534
pH	0.281	0.953
DO	0.995	0.035
BOD	−0.977	−0.280
COD	−0.978	−0.260
TSS	−0.102	0.952
NH$_3$N	−0.849	−0.410
Rainfall	0.679	−0.424
Velocity	−0.682	0.639
Epifaunal	0.968	0.057
Vegetative	0.905	−0.307
Eigenvalue	0.842	0.447
Percentage of variance explained	57.85	30.72

Table 6. Summary statistic for the canonical correspondence analysis (CCA) relating aquatic macroinvertebrate–environmental variables (11 variables).

7. Conclusion

Rivers in Tropical Asia is closely related to the effect of seasonal flow imposed by unpredictable monsoon and seasonal rainfall [34, 35]. Several studies have analyzed the changes of benthic community dictated by seasonal rainfall [36–38], including those in Peninsular Malaysia [39–41]. In general, rainfall pattern in Malaysia is much influenced by wind flow pattern during the seasonal period. To some extent, it is also influenced by local topography. The present study indicated that seasonal rainfall has not significantly affected the distribution of benthic communities in the four studied rivers. This study is in line with the findings from Refs. [33, 42]. The anthropogenic impacts were more significant than the seasonal rainfall. However, the EPT populations were seen correlated to seasonal variation, as been reported earlier by Suhaila et al. [40] in Gunung Jerai Forest Reserve. According to Ref. [43], these types of species react quickly to the changes in environment.

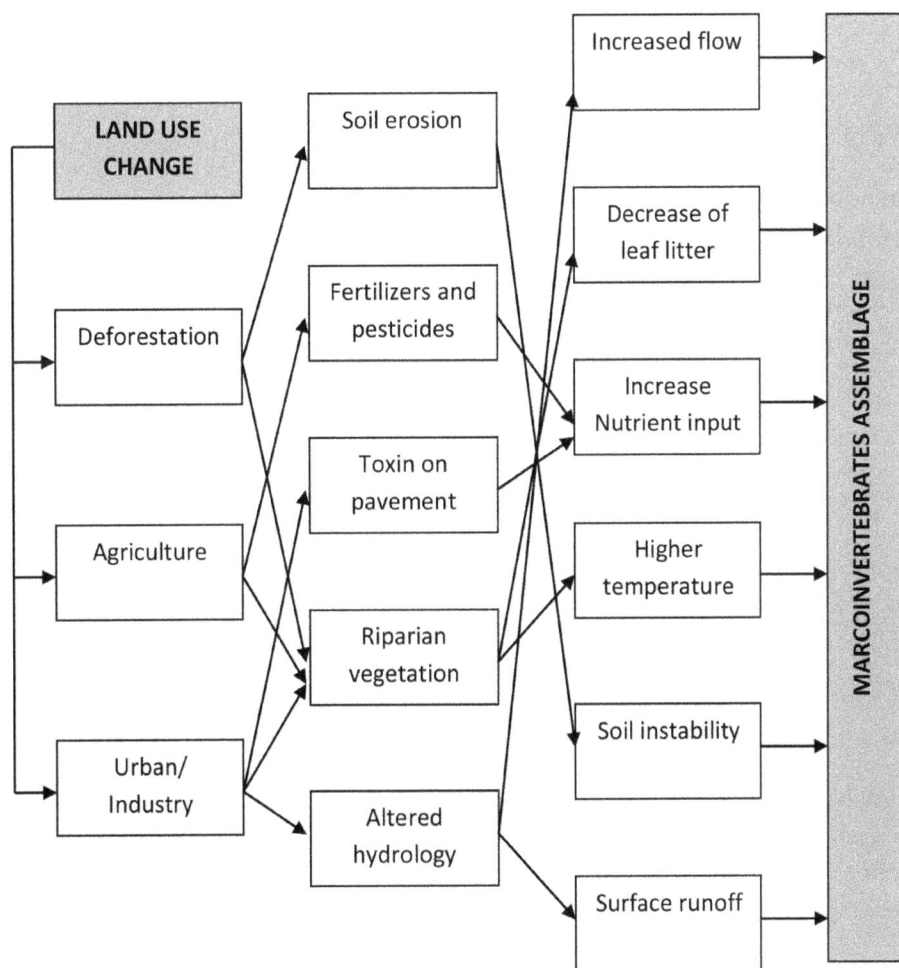

Figure 2. Illustrative schematic of the potential interactions between the threats, impact, and the response to the stream macroinvertebrate assemblage.

The present findings reveal that human-induced changes in natural habitat are the ultimate causes explaining the alteration in macroinvertebrates biodiversity. There is no doubt that anthropogenic disturbance impacted the structure of macroinvertebrate communities, either in the tropics or in the temperates [44–47]. Land use change is an integration of various human activities that negatively impact the river ecosystem. Among which, flow variability and sediments transport into the river by impervious surface and drainage in urban areas, stream channelization, agriculture, and deforestation. All of these threats will be manifested in changes in flows, benthic habitat conditions, and riffle-pool integrity. **Figure 2** presents an illustrative example of how macroinvertebrate communities can respond to land use change through a chain of indirect effects that lead to changes to the macroinvertebrate assemblage in both taxa richness and relative abundance.

Author details

Salmiati[1,2*], Nor Zaiha Arman[1] and Mohd Razman Salim[1,2]

*Address all correspondence to: salmiati@utm.my

1 Department of Environmental Engineering, Faculty of Civil Engineering, University of Technology, Johor Bahru, Johor, Malaysia

2 Center for Environmental Sustainability and Water Security (IPASA), RISE, University of Technology, Johor Bahru, Johor, Malaysia

References

[1] Nienhuis PH. Water and values: ecological research as the basis for water management and nature management. Hydrobiologia. 2006 Jul;565(1):261–75. doi:10.1007/s10750-005-1918-2.

[2] Macionis JJ, Parrillo VN. Cities and Urban Life. Upper Saddle River, NJ: Pearson Education; 2004.

[3] Chan Ngai Weng, Abdullah AL, Ibrahim AL and Ghazali S. River pollution and restoration towards sustainable water resources management in Malaysia. National Seminar on Society, Space and Environment in a Globalised World: Prospects & Challenges, 29–30 May 2003. The City Bayview Hotel, Penang, p. 208–219.

[4] Rahman HA. An Overview of River Pollution Issues in Malaysia. Retrieved on March. 2007;26:2014.

[5] Andaya BW, Andaya LY. A History of Malaysia. University of Hawaii Press; 2001.

[6] Mohamad S, Toriman ME, dan Mokhtar Jaafar KA. River and Development: Waterfront City Malaysia. Bangi: UKM; 2015.

[7] Department of Statistics Malaysia. Compendium of Environment Statistics Malaysia. Kuala Lumpur: Department of Statistics Malaysia; 2013.

[8] Harian S. Favourable Initiative of River Conservation (2013 September). Retrieved 26 May 2016, from http://www.sinarharian.com.my/article/conservation inisiative-river-favourable-1.198987.

[9] Gordon ND, Finlayson BL, McMahon TA. Stream Hydrology: An Introduction for Ecologists. John Wiley and Sons; 2004 Jun.

[10] Conti ME. Biological Monitoring: Theory & Applications: Bioindicators and Biomarkers for Environmental Quality and Human Exposure Assessment. WIT Press; 2008.

[11] Wei M, Zhang N, Zhang Y, Zheng B. Integrated assessment of river health based on water quality, aquatic life and physical habitat. Journal of Environmental Sciences. 2009 Dec;21(8):1017–1027. doi:10.1016/S1001-0742(08)62377-3

[12] Iliopoulou-Georgudaki J, Kantzaris V, Katharios P, Kaspiris P, Georgiadis T, Montesantou B. An application of different bioindicators for assessing water quality: a case study in the rivers Alfeios and Pineios (Peloponnisos, Greece). Ecological Indicators. 2003 Feb; 2(4):345–360. doi:10.1016/S1470-160X(03)00004-9

[13] Liebmann H. Manual of Freshwater and Sewage Biology. In Manual of Freshwater and Sewage Biology. R. Oldenbourg; 1962.

[14] Khan NISA. Assessment of water pollution using diatom community structure and species distribution—a case study in a tropical river basin. Internationale Revue der gesamten Hydrobiologie und Hydrographie. 1990 Jan;75(3):317–338. doi:10.1002/iroh.19900750305

[15] Arsad A, Abustan I, Rawi CS, Syafalni S. Integrating biological aspects into river water quality research in Malaysia: an opinion. OIDA International Journal of Sustainable Development. 2012 May;4(02):107–122.

[16] Ghani WM, Rawi CS, Hamid SA, Al-Shami SA. Efficiency of different sampling tools for aquatic macroinvertebrate collections in Malaysian streams. Tropical Life Sciences Research. 2016 Feb;27(1):115.

[17] Zaiha AN, Ismid MM, Salmiati, Azri MS. Effects of logging activities on ecological water quality indicators in the Berasau River, Johor, Malaysia. Environmental Monitoring and Assessment. 2015 Aug;187(8):1–9. doi:10.1007/s10661-015-4715-z

[18] Sharifah Aisyah SO, Aweng ER, Razak W, Ahmad Abas K. Preliminary study on benthic macroinvertebrates distribution and assemblages at Lata Meraung waterfall, Pahang, Malaysia. Jurnal Teknologi. 2015 Jan;72(5): 1–4

[19] Al-Shami SA, Rawi CS, Ahmad AH, Hamid SA, Nor SA. Influence of agricultural, industrial, and anthropogenic stresses on the distribution and diversity of macroin-

vertebrates in Juru River Basin, Penang, Malaysia. Ecotoxicology and Environmental Safety. 2011 Jul;74(5):1195–1202. doi:10.1016/j.ecoenv.2011.02.022

[20] Al-Shami SA, Rawi CS, HassanAhmad A, Nor SA. Distribution of Chironomidae (Insecta: Diptera) in polluted rivers of the Juru River Basin, Penang, Malaysia. Journal of Environmental Sciences. 2010 Nov;22(11):1718–1727. doi:10.1016/ S1001-0742(09)60311-19

[21] RM2 juta atasi masalah banjir di Mengkibol (2011 September). Utusan Online. Retrieved May 14, 2015, from http://www.utusan.com.my/utusan/info.asp? y=2011&dt=0923&pub=Utusan_Malaysia&sec=Johor&pg=wj_04.htm.

[22] Barbour MT, Verdonschot PF, Stribling JB. The multihabitat approach of USEPA's rapid bioassessment protocols: benthic macroinvertebrates. Limnetica. 2006;25(3):839–850.

[23] Yong HS, Yule CM. Freshwater Invertebrates of the Malaysian region. Akademi Sains Malaysia; 2004.

[24] Federation WE, American Public Health Association. Standard methods for the examination of water and wastewater. American Public Health Association (APHA): Washington, DC, USA; 2005.

[25] Quinn JM, Cooper AB, Davies-Colley RJ, Rutherford JC, Williamson RB. Land use effects on habitat, water quality, periphyton, and benthic invertebrates in Waikato, New Zealand, hill-country streams. New Zealand Journal of Marine and Freshwater Research. 1997 Dec;31(5):579–597. doi:10.1080/00288330.1997.9516791

[26] Triplehorn CA, Johnson NF. Study of Insect. Thompson Brooks: United States of America; 2005.

[27] Lancaster J, Downes BJ, Arnold A. Oviposition site selectivity of some stream-dwelling caddisflies. Hydrobiologia. 2010 Sep;652(1):165–178. doi:10.1007/s10750-010-0328-2

[28] Wowor D, Cai Y, Ng, PK. Crustacea: Decapoda, Caridea. In Yong HS, Yule CM (eds.). Freshwater Invertebrates of the Malaysian Region. Malaysian Academy of Sciences, 2004; p. 337–357.

[29] Wowor D, Muthu V, Meier R, Balke M, Cai Y, Ng PK. Evolution of life history traits in Asian freshwater prawns of the genus Macrobrachium (Crustacea: Decapoda: Palaemonidae) based on multilocus molecular phylogenetic analysis. Molecular Phylogenetics and Evolution. 2009 Aug;52(2):340–350. doi:10.1016/j.ympev.2009.01.002

[30] Yule CM. Insecta: Diptera. In Yule CM and Yong HS (eds.) Freshwater Invertebrates of the Malaysian Region. Academy of Sciences Malaysia, 2004; p. 610–612.

[31] Marziali L, Armanini DG, Cazzola M, Erba S, Toppi E, Buffagni A, Rossaro B. Responses of Chironomid larvae (Insecta, Diptera) to ecological quality in

Mediterranean river mesohabitats (South Italy). River Research and Applications. 2010 Oct;26(8):1036–1051. doi:10.1002/rra.1303

[32] Weber RE, Vinogradov SN. Nonvertebrate hemoglobins: functions and molecular adaptations. Physiological Reviews. 2001 Apr;81(2):569–628.

[33] Arman NZ, Salmiati, Said MI, Azman S. anthropogenic influences on aquatic life community and water quality status in Mengkibol River, Kluang, Johor, Malaysia. Journal of Applied Sciences in Environmental Sanitation. 2013 Sep;8(3):151–160.

[34] Dudgeon D. The ecology of tropical Asian rivers and streams in relation to biodiversity conservation. Annual Review of Ecology and Systematics. 2000 Jan;Vol. 31: 239–263.

[35] Gopal B. Conserving biodiversity in Asian wetlands: issues and approaches. The Asian Wetlands: Bringing Partnerships into Good Wetland Practices. Penerbit Universiti Sains Malaysia: Pulau Pinang, Malaysia; 2002; p. ID15–ID22.

[36] Silveira MP, Buss DF, Nessimian JL, Baptista DF. Spatial and temporal distribution of benthic macroinvertebrates in a southeastern Brazilian river. Brazilian Journal of Biology. 2006 May;66(2B):623–632. doi:10.1590/S1519-69842006000400006

[37] Joshi PC, Negi RK, Negi T. Seasonal variation in benthic macro-invertebrates and their correlation with the environmental variables in a freshwater stream in Garhwal region (India). Life Science Journal. 2007;4(4):85–89.

[38] Waite IR, Herlihy AT, Larsen DP, Urquhart NS, Klemm DJ. The effects of macroinvertebrate taxonomic resolution in large landscape bioassessments: an example from the Mid-Atlantic Highlands, USA. Freshwater Biology. 2004 Apr;49(4):474–489. doi: 10.1111/j.1365-2427.2004.01197.x

[39] Harun S, Al-Shami SA, Dambul R, Mohamed M, Abdullah MH. Water quality and aquatic insects study at the lower Kinabatangan River catchment, Sabah: in response to weak la niña event. Sains Malaysiana. 2015 Apr;44(4):545–558.

[40] Suhaila AH, Che Salmah MR, Nurul Huda A. Seasonal abundance and diversity of aquatic insects in rivers in Gunung Jerai Forest Reserve, Malaysia. Sains Malaysiana. 2014;43(5):667–674.

[41] Suhaila AH, Salmah MR, Al-Shami SA. Temporal distribution of Ephemeroptera, Plecoptera and Trichoptera (EPT) adults at a tropical forest stream: response to seasonal variations. The Environmentalist. 2012 Mar;32(1):28–34. doi:10.1007/s10669-011-9362-5

[42] Azrina MZ, Yap CK, Ismail AR, Ismail A, Tan SG. Anthropogenic impacts on the distribution and biodiversity of benthic macroinvertebrates and water quality of the Langat River, Peninsular Malaysia. Ecotoxicology and Environmental Safety. 2006 Jul; 64(3):337–347. doi:10.1016/j.ecoenv.2005.04.003

[43] Flecker A, Feifareck B. Disturbance and temporal variability of invertebrate assemblages in two Andean stream. Freshwater Biology. 1994;31:131–142. doi:10.1111/j. 1365-2427.1994.tb00847.x

[44] Liow LH, Sodhi NS, Elmqvist T. Bee diversity along a disturbance gradient in tropical lowland forests of south-east Asia. Journal of Applied Ecology. 2001 Feb;38(1):180–192. doi:10.1046/j.1365-2664.2001.00582.x

[45] Hanski I, Koivulehto H, Cameron A, Rahagalala P. Deforestation and apparent extinctions of endemic forest beetles in Madagascar. Biology Letters. 2007 Jun;3(3):344–347. doi:10.1098/rsbl.2007.0043

[46] Sodhi NS, Lee TM, Koh LP, Brook BW. A meta-analysis of the impact of anthropogenic forest disturbance on Southeast Asia's Biotas. Biotropica. 2009 Jan;41(1):103–109. doi: 10.1016/j.tree.2004.09.006

[47] Sandin L, Solimini AG. Freshwater ecosystem structure–function relationships: from theory to application. Freshwater Biology. 2009 Oct;54(10):2017–2024. doi:10.1111/j. 1365-2427.2009.02313.x

4

Assessment of Impacts of Acid Mine Drainage on Surface Water Quality of Tweelopiespruit Micro-Catchment, Limpopo Basin

Bloodless Dzwairo and Munyaradzi Mujuru

Additional information is available at the end of the chapter

Abstract

This research aimed to contribute to current literature for Tweelopiespruit micro-catchment, Limpopo Basin, by trending SO_4^{2-}, Cl^-, Ca^{2+}, Mg^{2+}, Na^+, K^+, Fe, pH and EC, for points F1S1, F2S2, W1S3, F6S7, F8S9, F10S11 and F11S12, as identified by the Department of Water and Sanitation, South Africa, for years 2003 to 2008. Results showed that pollutant concentrations generally increased downstream, which questioned their possible sources since pollution generally attenuates towards downstream. A possible explanation was that groundwater (polluted with the effluent) could be decanting from various places, thus contributing to the increase in concentrations, in places. This could potentially add value to existing efforts, which aim to halt and reverse impacts of acid mine drainage (AMD) in the micro-catchment and possibly in the Goldfields (a highly negatively impacted environment), which incorporates the Cradle of Humankind. Conclusions reached could provide invaluable options for alternative technological or methodological approaches that could be adopted for the treatment of AMD. This is critical to South Africa's water quality trending and sustainability of this ecosystem, especially because the Tweelopiespruit micro-catchment supports humans and a variety of wildlife like giraffe, within the preserve of the Krugersdorp Game Reserve (KGR) and also its outer boundaries.

Keywords: acid mine drainage, Limpopo Basin, Tweelopiespruit micro-catchment, water quality, pollution

1. Introduction

Acid mine drainage (AMD) is a pollutant that arises from exposure of metal sulphide minerals such as the abundantly available pyrite (FeS_2) to oxygen and water during the mining of metals and coals [1, 2]. Pyrite undergoes oxidation in a series of reactions, the first stage (trigger) of which results in production of sulfuric acid and ferrous sulfate as provided in

Eq. (1) [2]. The last stage results in formation of stable and soluble ferric iron (at pH lower than 3.5) or formation of the red precipitate ferric hydroxide (at pH greater than 3.5) [2]. Although AMD formation processes are accelerated by exposure to air [1], in oxygen-independent reactions, ferric iron becomes the main oxidant of the various other metal sulfides, which tend to associate with the pyrite in mineral formations. Naturally occurring bacteria can speed up the formation of AMD when they break down sulfide minerals [3]:

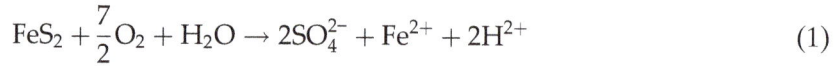

$$FeS_2 + \frac{7}{2}O_2 + H_2O \rightarrow 2SO_4^{2-} + Fe^{2+} + 2H^{2+} \tag{1}$$

Because pyrite is associated with gold and coal formations, mining of these minerals has subsequently resulted in very toxic and degraded environment, which are mainly highly acidic and usually contain excessive concentrations of metals, sulfides, sulfates, heavy metals, and salts [2, 4–9]. This is noted even in the South African content where it has been shown that coal formations of the Permian and Triassic-Permian ages, which lie in the E. Kalahari Precambrian Belt and the formations of the Permian, Permian–Carboniferous, and Triassic ages found in the Karoo Supergroup, are associated with gold deposits (**Figure 1** [10]). Indeed, this coformation means that large tracts of the South African environment are impacted by AMD.

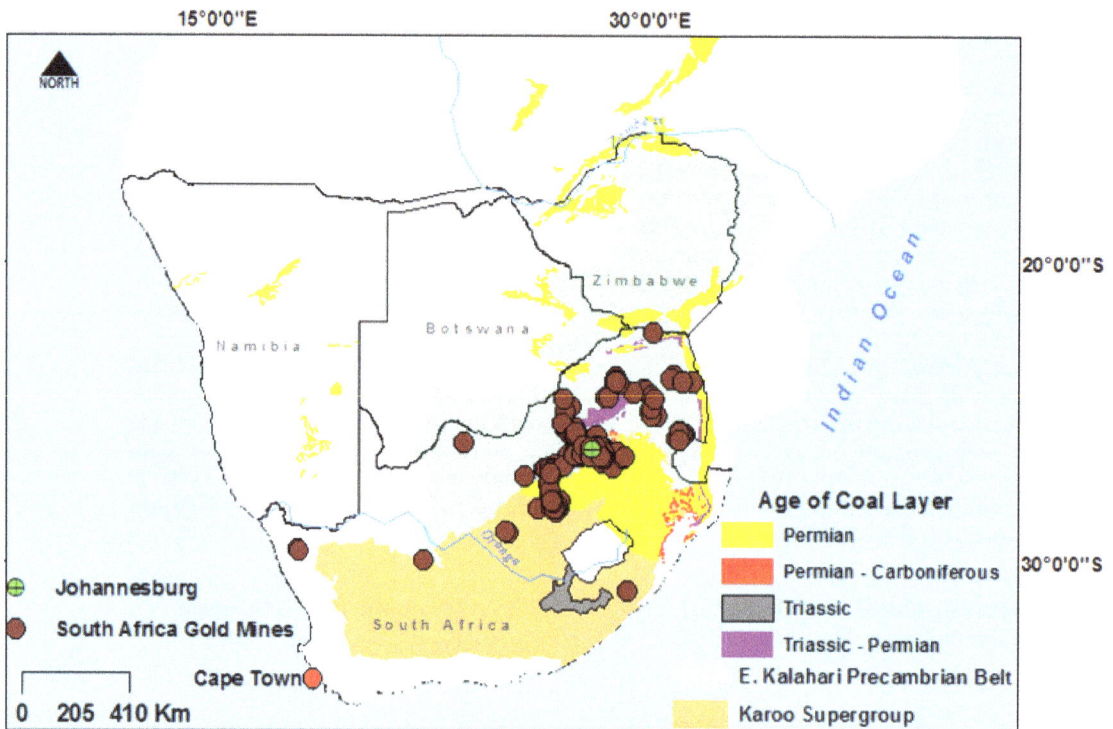

Figure 1. South Africa's gold mine locations and coal deposits (Software platform: ESRI [10]. Source of shapefiles: Internet).

At a global level, the latest Blacksmith's report by Harris and Andrew [11] has provided a tool in the form of a geospatially coded map (**Figure 2** [11]), to assist governments with prioritizing future resource allocation and pollution clean-up efforts. In this report it has been noted that mining activities occupy positions number one (artisanal gold mining), six and seven (mining and ore processing) in the top 10 of the world's 20 worst toxic pollution problems [11]. All three activities aforementioned are major sources of AMD, which Benedetto de Almeida [7] also described as one of the most serious environmental problems that the mining industry has ever created.

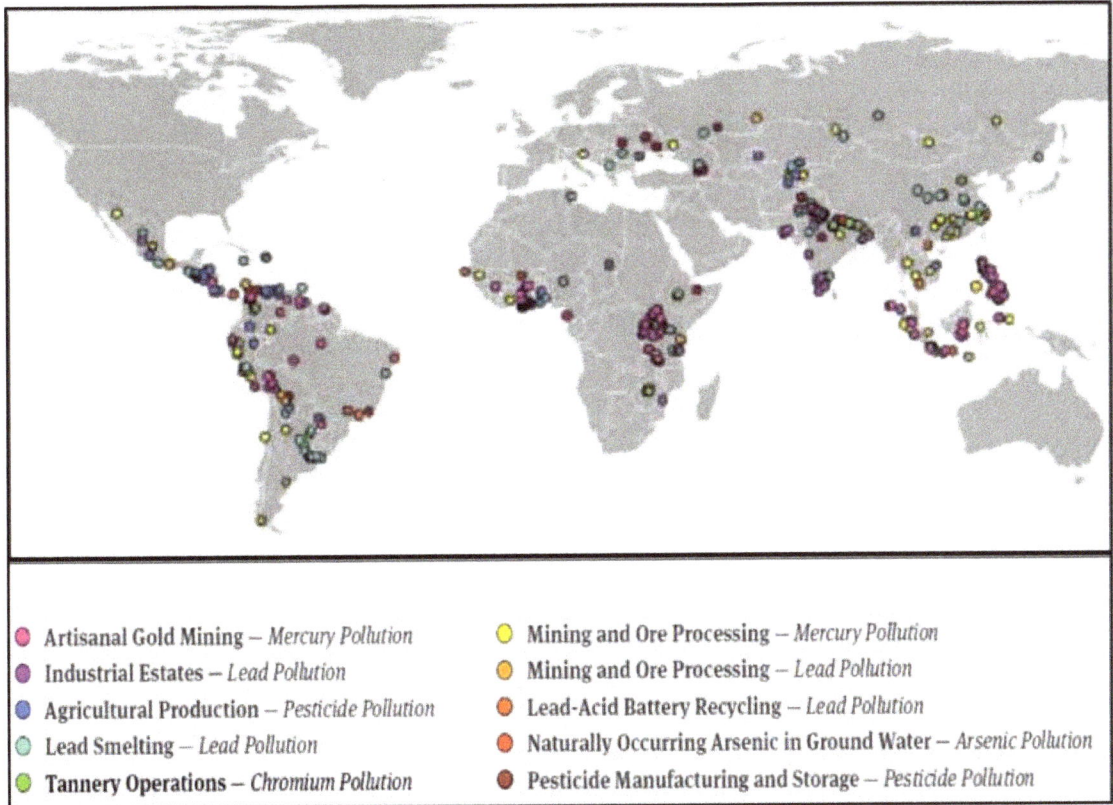

Artisanal Gold Mining – *Mercury Pollution* Mining and Ore Processing – *Mercury Pollution*
Industrial Estates – *Lead Pollution* Mining and Ore Processing – *Lead Pollution*
Agricultural Production – *Pesticide Pollution* Lead-Acid Battery Recycling – *Lead Pollution*
Lead Smelting – *Lead Pollution* Naturally Occurring Arsenic in Ground Water – *Arsenic Pollution*
Tannery Operations – *Chromium Pollution* Pesticide Manufacturing and Storage – *Pesticide Pollution*

Figure 2. Geospatially coded map of top ten of the world's 20 worst toxic pollution processes [12].

Although Harris and Andrew [11] highlights that the South African environment is impacted by pesticide residues, Zilles Peccia [13] argued that AMD is the single most significant threat to the country's environment. For example, other researchers concurred that apart from the fact that mine dumps create harsh acidic and chemically toxic ecosystems in the country, a major environmental concern of pollution from AMD is the severe impact it has on productive land (e.g., agricultural land) as well as on groundwater, surface water, and aquatic life (e.g., the Vaal River Basin) as shown in (**Figure 3** [10, 14, 15]).

Therefore, treating AMD-impacted environments is a priority for South Africa as much as it is for the world, because if the environments are left as they are, the problem will just get worse, rendering more and more ecosystems uninhabitable. Evidently there are large tracts of land in South Africa, which are unusable because they are already impacted by AMD, examples having been documented in the East, Central, and Western basins of South Africa's Goldfields.

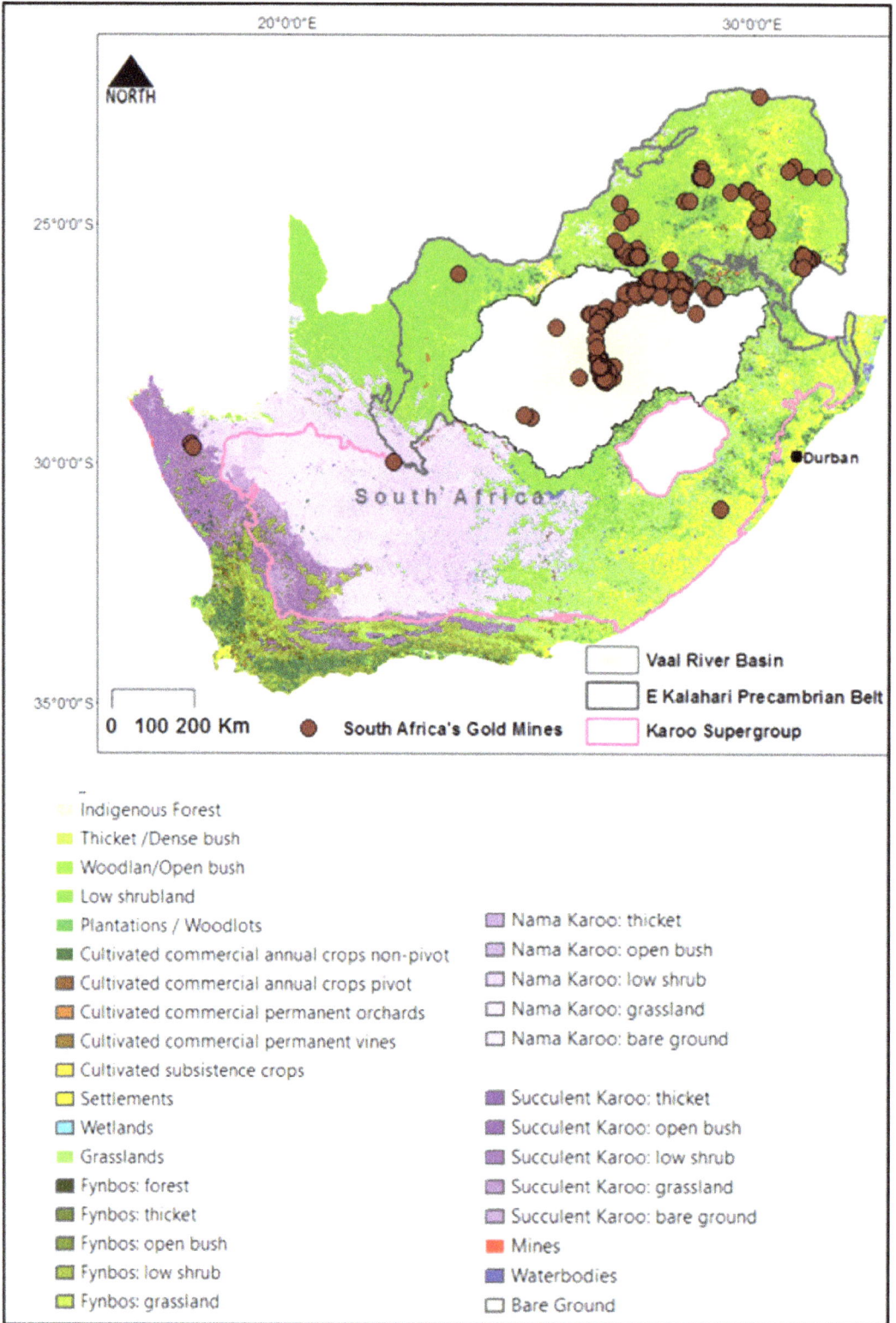

Figure 3. South Africa's land cover and the locations of gold mines (Software platform: ESRI [10]. Source of shapefiles: Internet).

Here, surface and groundwater are extremely polluted and unusable [11, 15–17] because gold and coal are mined largely from ores that also contain pyrite. The underlying hard-rock unit is made up of the Witwatersrand System in combination with others like the Transvaal System-Dolomite, The Ecca System, The Karoo, etc. It is noted that the Witwatersrand System, as represented by the Witwatersrand mines, is completely located in the Vaal Basin, a very strategic basin in South Africa, as indicated in **Figure 4** [10]. Therefore, due to the economic implications of polluting key livelihood environment, it has been suggested that where treatment processes are economically feasible and practical, it is necessary to reclaim the impacted environment and mitigate against pollution.

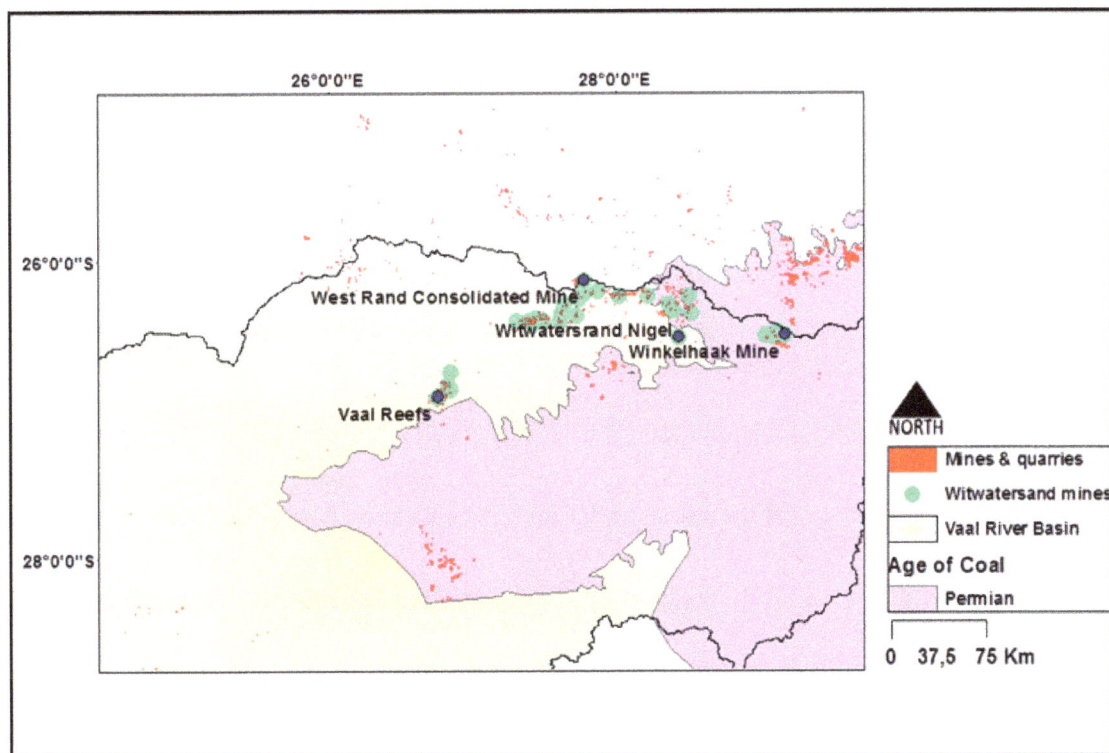

Figure 4. Witwatersrand System of gold mines located entirely in the Vaal Basin (Software platform: ESRI [10]. Source of shapefiles: Internet).

For example, Kruse, Bowman [15] reported on AMD treatment in a watershed near the village of Carbondale, Ohio, Hewett Fork subwatershed. The treatment process utilized involved neutralizing the AMD with lime. Results indicated that between the years 2000 and 2004, pH had improved from about four to around nine, with concomitant improvement in the biological communities in the study area. The major conclusion drawn from this intervention was that a 2-week interruption in treatment impacted on the fish community to a great extent while the macroinvertebrate community showed very minor perturbation. The reported community shift is a typical phenomenon for perturbed trophic structures [15].

Additionally, Wei, Wei [14] conducted a stream monitoring study in the United States for a period of 7 years. The objective was to evaluate the water quality trend and land cover in a

Mid-Appalachian watershed. The study area was a reclaimed former coal mining environment. GIS tools and multivariate analysis were applied to correlate the water quality trends and land cover. Results for pH, sulfate, and metals indicated that AMD was the major factor leading to overall poor water quality. It was concluded that water quality improvement was evident in subwatersheds which were originally heavily impacted but which were later reclaimed by reforestation. This indicated that good reclamation practices had positive impacts on water quality over time [14].

Benner, Gould [4] instead used bacterial populations and water chemistry to profile groundwater at Nickel Rim mine tailings impoundment in Ontario, Canada. The objective was to trace a plume of pollutants from the tailings impoundment and to find out if that plume was impacting groundwater in the vicinity. Results from groundwater analysis showed elevated populations of iron and sulfur oxidizing bacteria. These bacterial populations were restricted to hydrologically defined zones of recharge and discharge. It was concluded that active oxidation in the Nickel Rim tailings was occurring immediately above the water table, where water content was high in comparison to unsaturated zones further away from the water table. One plausible reason for this was that the water table interface provided continuous moisture gradient/potential difference enough to sustain ideal conditions for bacterial growth [4].

Despite these best efforts to try and reclaim impacted ecosystems, legal instruments have fallen short of implementing recommendations in order to deter further environmentally insensitive mining activities, challenges abound. For example, in South Africa, it has not been able to offer legal recourse for mine-related polluted environment because the situation is very complex, even though South Africa's Constitution [18] and related legal instruments support environmental sustainability. Many of the mines are closed off or dysfunctional, which should call for directors of these former mines to be answerable [19–22] for prosecution or jail terms, yet destruction continues. Implementation of the legal instruments seems to be the major stumbling block.

As part of on-going technology trials in South Africa's Witwatersrand Goldfields, Bologo, Maree [23] conducted experiments in order to understand the dynamics of reducing concentrations of Ca, Fe, SO_4^{2-}, and Mg from AMD-polluted effluent. The magnesium-barium-oxide process resulted in a reduction of pollutant concentrations. The technology also managed to recover the starter chemicals for reuse [23].

De Beer, Maree [24] used a CSIR ABC desalination process in a pilot plant to neutralise AMD samples from the Western Basin of the Goldfields. The process managed to remove total dissolved solids from 2600 to 360 mg/L. Metals were precipitated with CaS, Ca(HS)$_2$, or Ca (OH)$_2$ while SO_4^{2-} was reduced to 100 mg/L in a two-step process that employed gypsum crystallization followed by BaCO$_3$ treatment. The starting raw materials were recovered for reuse, making the process sustainable and cost-effective. It was demonstrated that the final treated water met the South African National Standard (SANS) 241 drinking water quality standard [24]. Motaung, Maree [25] used the same pilot plant to demonstrate that South Africa could produce sulfur from AMD treatment at a cost of ZAR2.21 m^{-3} of raw effluent. The potential value of the water and by-products amounted to R11.10 m^{-3} at a Rand value of US $1.00 = ZAR7.60 [25].

To support rehabilitation efforts, many studies have been done with the aim of characterizing and/or analyzing AMD as well as assessing its impact on ecosystems. Whichever rehabilitation process is chosen; the resultant treated effluent should be of good quality that is fit-for-purpose. Treated water may be channeled back into the mining operations or it could be released into the natural environment, while precautions should be taken not to transfer pollution from one stream to another. The receiving environment should be able to recover but if this is not possible, the effluent could be diverted elsewhere. Strict monitoring and evaluation of the effluent (treated or raw) could form part of the strategic long-term planning when mitigating against AMD impacts [17].

It has been documented that AMD has seriously impacted the surface water quality of the Eastern, Central and Western basins of the Witwatersrand and Goldfields [2, 9, 17, 24–29]. Because South Africa relies heavily on surface water for drinking and agricultural purposes, AMD thus threatens livelihoods of many as well as national economic returns from agriculture. Consequently, AMD impacts are expected to persist for the next centuries in a "do-nothing" scenario [17], which is unacceptable because while water quality is threatened directly, decanting AMD effluent also threatens to drown sensitive historical and wildlife sanctuaries around the City of Johannesburg Locus.

Researchers in South Africa and elsewhere, however, are continuously developing alternative interventions that require integrated implementation of a range of measures [5, 23–25, 30, 31] including neutralization, crystallization, and diversion (pumping the decanting effluent for reuse) in order to mitigate and rehabilitate affected environment. Various other example successes are reported for mitigating and treating AMD from polluted environment [30, 32–36]. However, these mitigatory activities impact on the environment and thus require monitoring in order to evaluate effectiveness of interventions.

Active and defunct gold and coal mines continue to pollute ecosystems through AMD and deposit of elements like radioactive material and heavy metals. First, the pollutant's acidity leads to a decrease in pH of the recipient water, should that water body have insufficient buffering capacity. Secondly, when pH in receiving water is lowered, some of the metals remain in soluble toxic form, thus making AMD a potent effluent for receiving watercourses [8].

In South Africa, although gold mining in the Witwatersrand System (see **Figure 5** [10]), is declining, massive closure in the 1990s caught the government unprepared for the environmental degradation, especially the rising of groundwater as it filled the voids, which were abandoned after mining activities had removed much of the precious element–bearing rocks.

Reactions of water exposed to pyrite and oxygen then subsequently created AMD whose postclosure decant is currently an enormous threat to the environment. Consequently, pollution could get worse if remedial activities are delayed or not implemented [17]. Additionally, polluted effluent from the mines and quarries that extend into the Limpopo Basin (**Figure 6** [10]) threatens to flood downstream environment including the Cradle of Humankind which continue to pollute ecosystems through AMD and speciation [9, 23, 37].

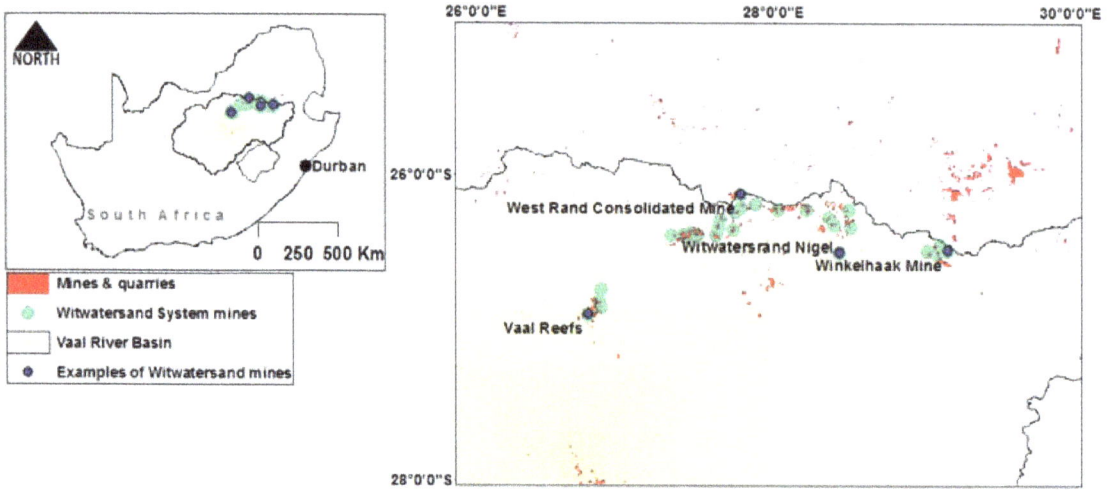

Figure 5. Witwatersrand System of gold mines in the Vaal Basin, South Africa (Software platform: ESRI [10]. Source of shapefiles: Internet).

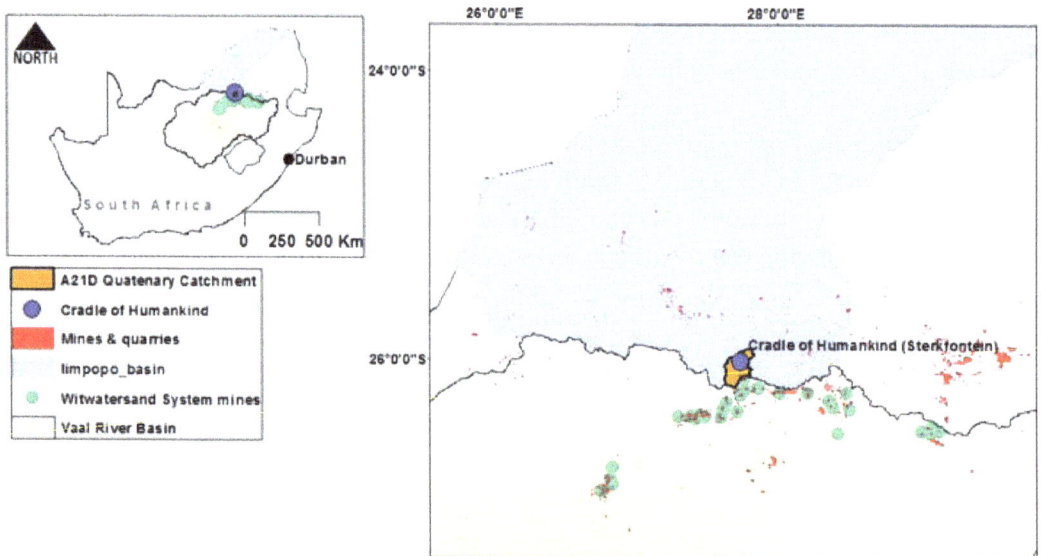

Figure 6. Threatened environments of the Cradle of Humankind, South Africa (Software platform: ESRI [10]. Source of shapefiles: Internet).

The Cradle of Humankind, which is located in the quaternary catchment of the Limpopo Basin called A21D, is one of South Africa's eight heritage sites and pollution threats of this magnitude are worrisome. Current initiatives are underway to either clean up the AMD before it reaches these vital and sensitive communities or stabilize it for reuse in fit-for-purpose situations [37–40]. These measures are crucial and strategic for the polluted environment where in 2010 about 60 ML/d AMD was decanting in the rainy season against a typical 20 ML/d in normal weather [37]. Decant polluted water flows via Tweelopiespruit and surrounding tributaries, through the Krugersdorp Game Reserve (KGR) and into Bloubankspruit that passes by the Cradle of Humankind, threatening this national treasure (**Figure 7** [10]). The KGR is home to a variety of wild animals which drink water from the polluted Tweelopiespruit.

To this end, solution development at Randfontein (**Figure 7** [10]) treatment plants include minimising the impact of waste (including AMD) from mining/AMD treatment, on the receiving aquatic environment by treating a portion of the effluent for re-use and release.

Using the Witwatersrand System effluent alone, researchers found out that the financial potential return of treating AMD was estimated at 350 ML/day (1ML = 1000 m³) [29]. This calculation revealed that if the effluent was treated back to raw water quality guidelines, it could represent 10% of the daily potable water supplied by Rand Water Board to municipalities in Gauteng Province and surrounding areas, at a cost of R3000/ML, indeed a financial justification to treat the polluted effluent from these environments.

Concomitantly, the current paper reports on research that aimed to contribute to research literature for the Tweelopiespruit, Limpopo Basin, South Africa, by assessing impacts of treated effluent on the Tweelopiespruit micro-catchment as a receiving environment. This was envisaged to enhance understanding of the extent of the AMD problem in order to inform on possible mitigation measures in the quaternary catchment A21D and possibly in the wider Vaal and Limpopo hydrological primary basins, which are the basins that are majorly impacted by gold and coal mining activities.

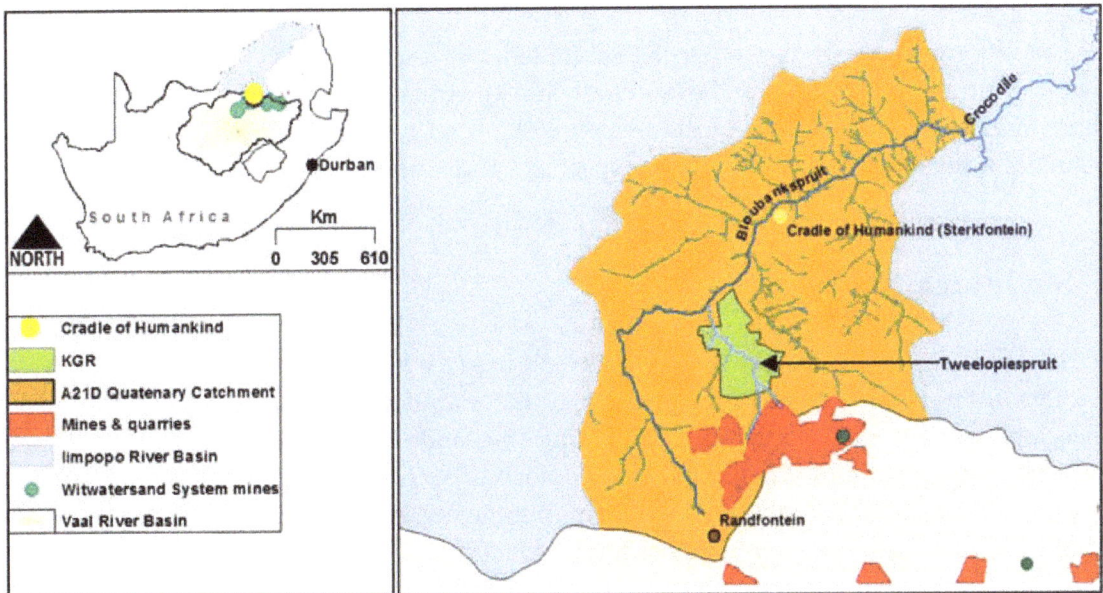

Figure 7. Tweelopiespruit, KGR and the Cradle of Humankind, South Africa (Software platform: ESRI [10]. Source of shapefiles: Internet).

Conclusions reached could provide information regarding whether treatment specifications could justifiably be continued as they were or could be enhanced to produce better quality treated effluent, especially as the Tweelopiespruit supports wildlife in the Krugersdorp Game Reserve (KGR).

2. Research problems and objectives

This research sought to answer the following questions:

- What are the trends in water quality of Tweelopiespruit and selected sampling sites around Randfontein plant, using the parameters SO_4^{2-}, Cl^-, Ca^{2+}, Mg^{2+}, Na^+ and K^+, Fe, pH, and electrical conductivity (EC)?

- What are the characteristics of the different water sample sources?

- What is the overall downstream water quality impact of the treatment plant intervention?

The overall aim of the research was to assess the impacts of treated AMD effluent on Tweelopiespruit's receiving and downstream ecosystem.

Specific objectives were:

- To evaluate the quality of water in the study area by analyzing and trending for SO_4^{2-}, Cl^-, Ca^{2+}, Mg^{2+}, Na^+, K^+, Fe, pH, and electrical conductivity (EC).

- To assess the impact of intervention (treatment plants) on Tweelopiespruit's health using the spatial and temporal trending patterns of the parameters.

The results from this study were to be submitted to the Team which was carrying out the overall neutralization process at the Randfontein AMD treatment plant. The report could assist them in assessing the impacts of the treated effluent on receiving waters, judging from the resultant water quality samples from the specified study monitoring sites.

3. Study area

In the Upper reaches of the Limpopo Basin's land use (see **Figure 8** [10]), polluted mine water decants from underground and flows from the Randfontein mining environment into Tweelopiespruit stream and the surrounding farming lands [17, 37, 39]. The massive discharge has altered the nature of this water course. Bologo et al. [21] estimated that about 50 ML is decanted into the Randfontein receiving environment each day. Some effluent enters the Tweelopiespruit wetland on the mine grounds via surface seepage [39].

Previous studies as reported by [25] identified the quality of water in the receiving karst groundwater environment as comprising a mixture of acid mine drainage and treated waste-water. This has compromised the sustainability of biodiversity (both plants and animals) in the KGR. This degeneration process currently (2013) poses a threat to the Cradle of Humankind, which has a geological formation of dolomitic rocks that are susceptible to attack and dissolution by AMD.

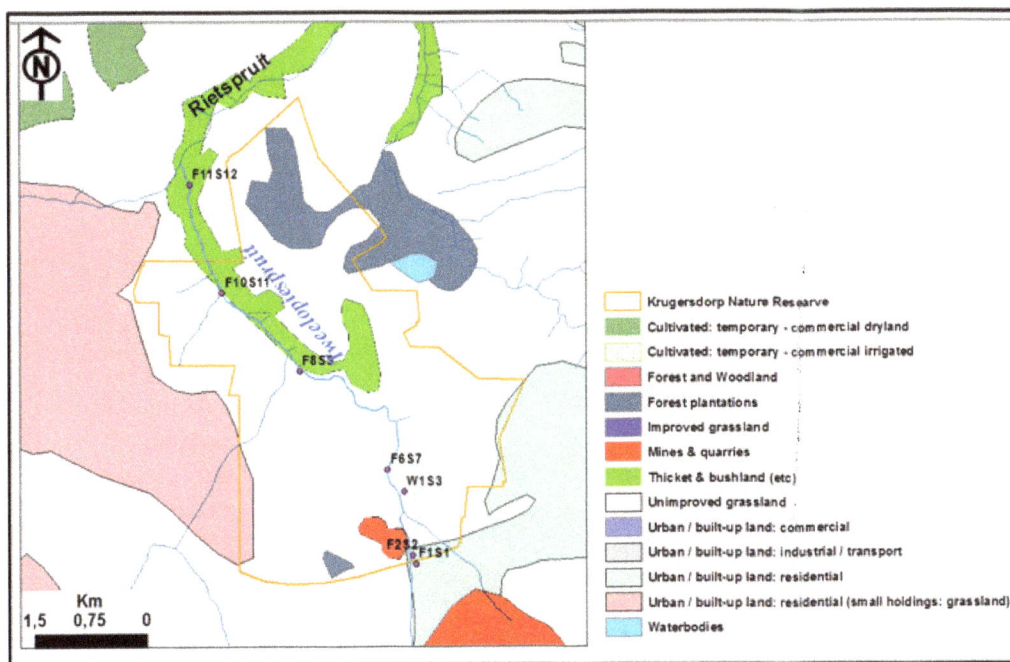

Figure 8. Tweelopiespruit land use (Software platform: ESRI [10]. Source of shapefiles: Internet).

Tweelopiespruit, which carries the sampling sites for this study, is a defined stream network thus it is identifiable as a hydrological unit and the seven sampling sites are clearly marked in **Figure 9**. Except for F1S1 and F11S12, five of the monitoring sites are located inside the KGR.

During the period of the study, the stream received both treated effluent AMD from three treatment pilot plants located at the Randfontein experimental site, as well as AMD that was decanting from underground within the surrounding environs. At one of the pilot treatment plants, AMD tertiary treatment was being employed, where a fraction of its volume (about 50% of decant volume by three treatment plants), was collected and treated in a dedicated neutralization plant that used lime $(Ca(OH)_2)$ [35]. Two of the three pilot plants that treated AMD used this method while the third one employed a different treatment technology. The lime process was reported by Khorasanipour, Moore [41] as a preferred method to others like the alkaline method, claiming that it had a high removal efficiency for dissolved heavy metals, relatively low cost, and was insensitive to seasonal temperature fluctuations. In all the three treatment plants, AMD was pumped from underground shafts to the surface for treatment before the treated effluent was released to Tweelopiespruit. It was expected that by pumping and treating the AMD, the underground level of the mine waste water would fall below a critical level to allow stoppage of the decantation process.

For this research, the sampling points were chosen to represent the flow of effluent and treated water within the micro-catchment. This allowed for trending based on spatial locations of the sampling sites as well as temporal and spatial analysis of the data. **Table 1** describes the sampling sites using the same identification which is used by the Department of Water and Sanitation (DWS).

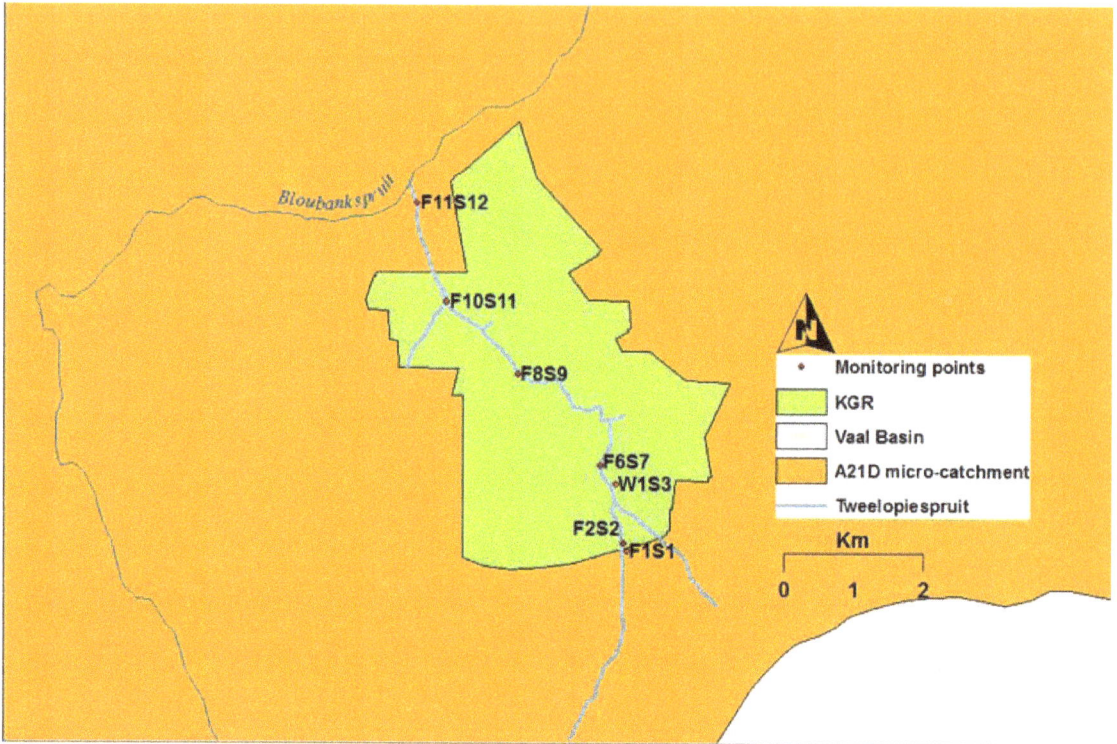

Figure 9. Sampling points along Tweelopiespruit (Software platform: ESRI [10]. Source of shapefiles: Internet).

Sampling point	Description	Latitude	Longitude
F1S1	Upstream of R24 at Randfontein Estates on Tweelopie	−26.10752	27.72268
F2S2	Willow tree in KGR on Tweelopie	−26.10653	27.72227
W1S3	Hippo Dam in KGR on Tweelopie	−26.09917	27.72128
F6S7	Cemetery Spring (1) (Spring 1) in KGR on Tweelopie	−26.09671	27.71932
F8S9	Lodge Spring (2) (Spring 2) at broad crest in KGR on Tweelopie	−26.08527	27.70886
F10S11	Northern fence in KGR	−26.07620	27.69963
F11S12	Tweelopie at the N14 intersection	−26.06374	27.69589

Table 1. Sampling sites on Tweelopiespruit.

4. Methods and materials

In this research, the quantitative research design was used. The research methods combined statistical analysis of retrospective (historical) data and batch analysis of water samples from the sites. Experimental analysis was performed on two batches of water samples, one in August and the other in September. Analysis was performed using the same analytical and standard methods which were used for the historical data in order to validate the historical data range that was employed in this study. The approach was flexible while mindful of the monitoring data which spanned hydrological years under a wide spectrum of hydrologic variability.

The experimental design had a random assignment where all samples from the monitoring sites had an equal chance of being assigned to a given experimental condition. Random assignment was used to ensure that experimental conditions did not differ significantly from each other.

An analytical approach was taken in order to determine the quality of water from the seven sampling points. The spatial and temporal changes were documented using tools that could aid in understanding the chemistry and the sources of the water, including geospatial mapping, and assessment of the impact of the treated effluent on the receiving water body.

- First, retrospective (historical) data were acquired from the DWS, which is a government entity that monitors chemical and biological quality of water as part of its mandate to provide environmental assessments and protection. It was treated for outliers before trending and geospatial mapping on ESRI ArcGIS 10.2.

- The second task involved conducting two sampling runs parallel with the regular monitoring exercise in the study area, in order to validate the retrospective data. The objective was to understand various aspects of the sampling sites in relation to the impacts of treated effluent on the quality of the receiving waters. Sampling and testing of water conformed to procedures for both the sampling and analysis of chemical parameters. All parameters were analyzed according to the standard methods of analysis.

5. Results and discussion

The results for chosen parameters, i.e., SO_4^{2-} (mg/L), Cl^- (mg/L), Ca^{2+} (mg/L), Mg^{2+} (mg/L), Na^+ (mg/L), K^+ (mg/L), Fe (mg/L), pH (pH units), and EC (m Sm^{-1}) are shown for all monitoring points in **Figures 10–20**; using trends that were plotted on MS Excel, for each sampling site according to their geocoordinates, starting with F11S12. While monitoring continues and is active at these sites since the 1970s, it is noted that there is a general trend toward increase in parameter concentration. For an environmentally sensitive micro-catchment, which also houses the Cradle of Humankind in its downstream ecosystem, efforts should be done to reduce the study area parameter concentrations in order improve the ecosystem. Because these are combination graphs for a mixture of parameters, individual units of measure could not be indicated on the y-axis of the graphs, hence the use of the label"parameter test value."

pH trends (**Figure 17**) for all monitoring points are lowest at the highest point (F1S1) but also is very unstable on all the other points because the environment itself is prone to nonpoint AMD pollution.

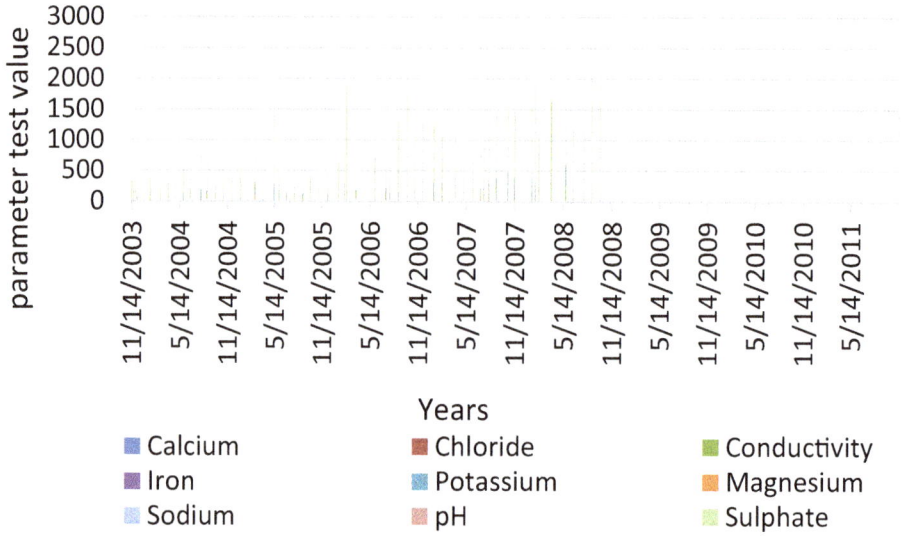

Figure 10. F11S12 monitored parameters for 2003–2011.

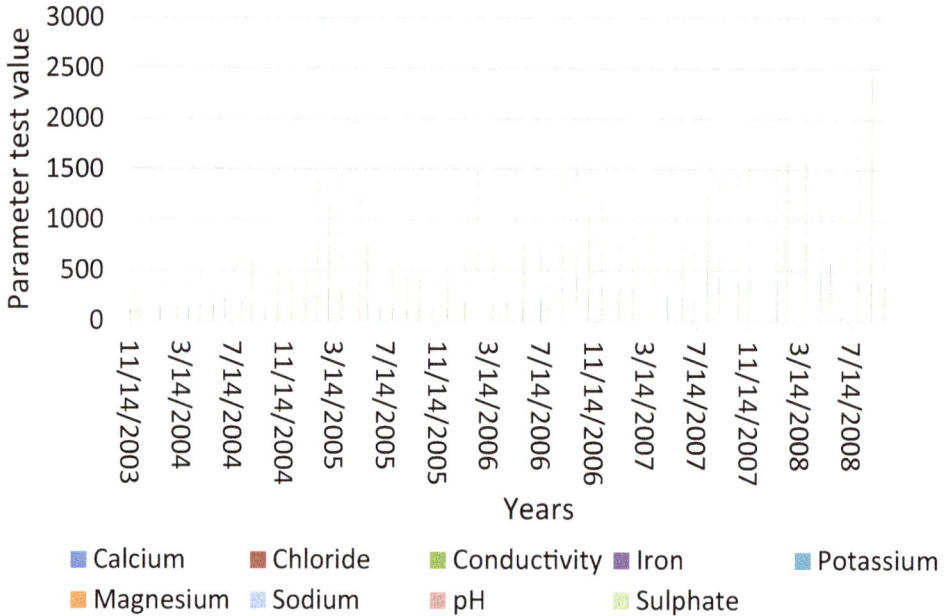

Figure 11. F10S11 monitored parameters for 2003–2008.

Just like the spiking values for pH, **Figure 18** indicates variable trends for sulfate along Tweelopiespruit.

However, the figure also indicates the conservative nature of sulfate as it shows lowest concentration at the sampling point farthest from the AMD sources along the river (F11S12 and

F10S11) and highest trends close to old mine shaft activities (F1S1 and F2S2). Sulfate does not mobilize easily and is therefore deposited along the stream soon after its entry. **Figure 19** shows that iron is typically high at the upper reaches of the river (F1S1), indicating the major source of AMD and a potential priority point for mitigation and management efforts to control AMD in the micro-catchment.

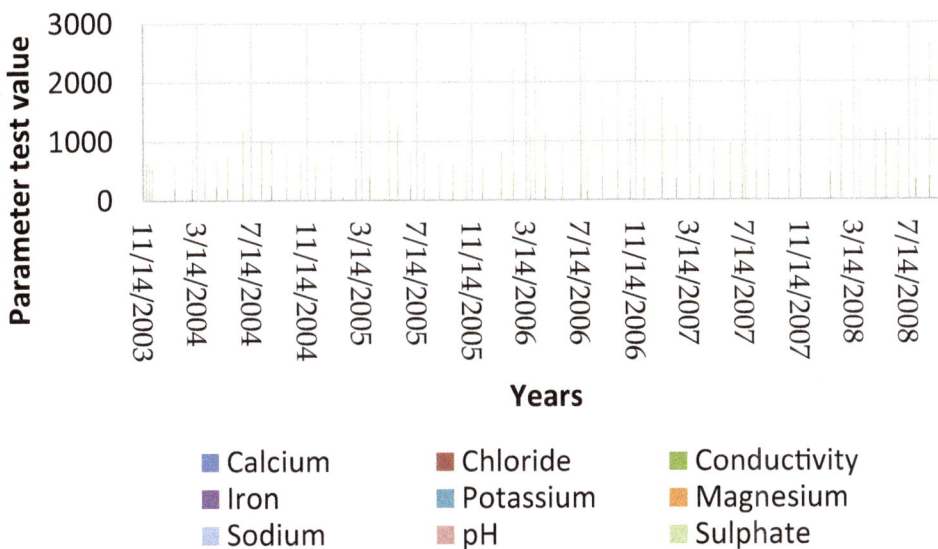

Figure 12. F8S9 monitored parameters for 2003–2008.

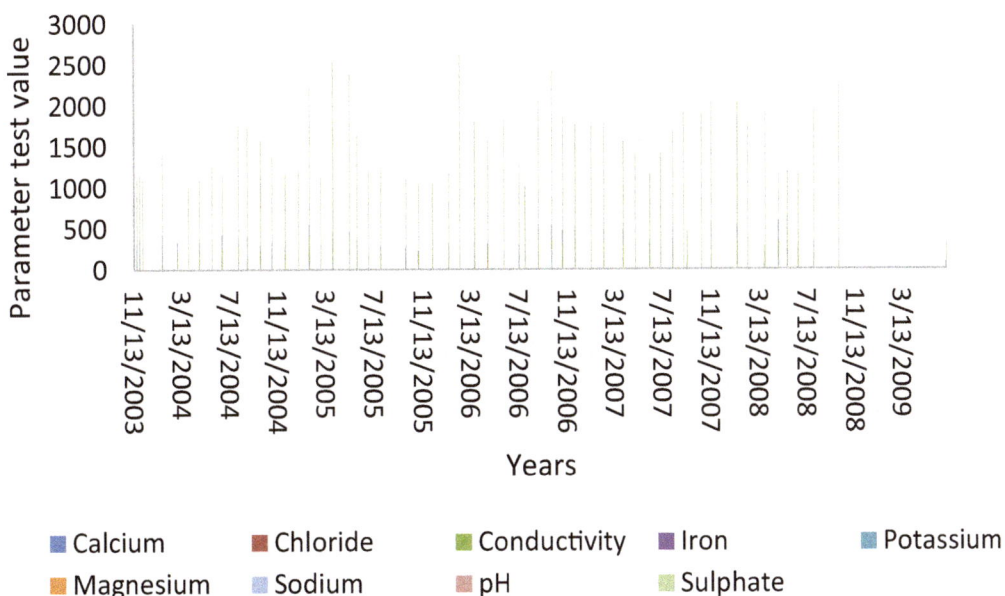

Figure 13. F6S7 monitored parameters for 2003–2009.

Calcium indicates a high concentration in the upper reaches of the river, too, as indicated in **Figure 20**.

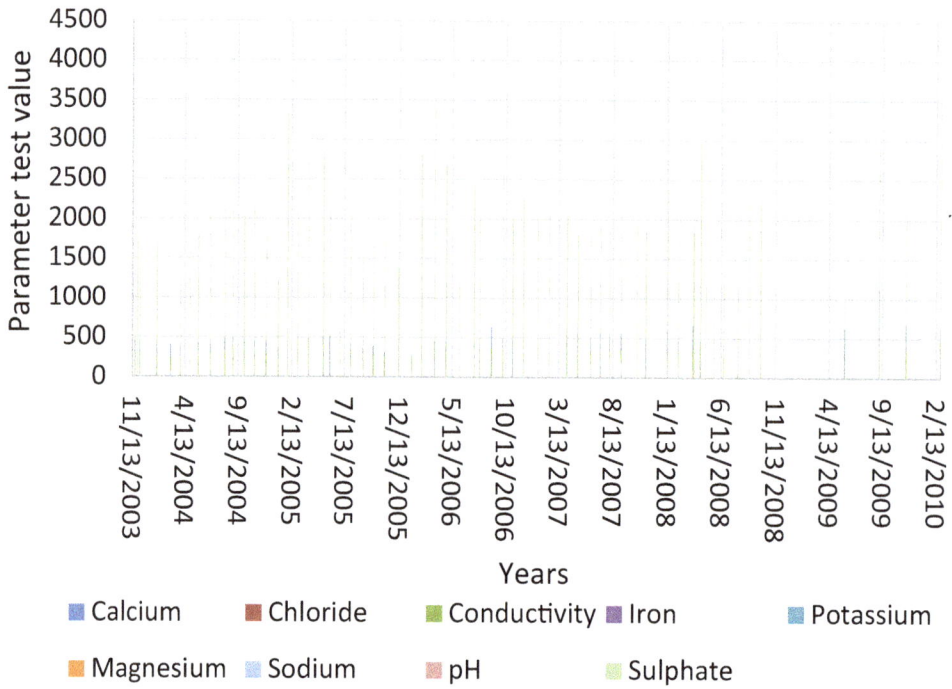

Figure 14. W1S3 monitored parameters for 2003–2010.

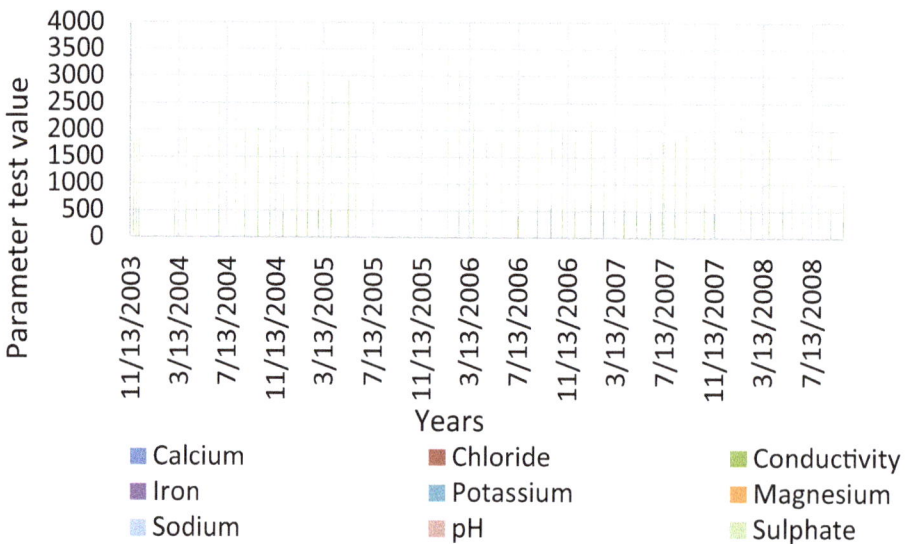

Figure 15. F2S2 monitored parameters for 2003–2008.

W1S3 indicates higher values for calcium and could be subject for further investigation regarding the type of water that passes by that monitoring point. For example, calcium contributes to water hardness which affects mobility of other related ions and anions in the water and sediments.

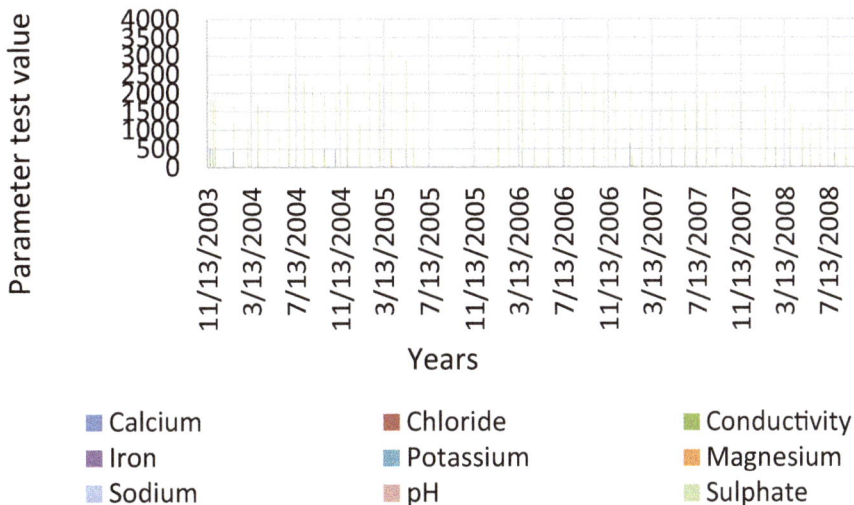

Figure 16. F1S1 monitored parameters for 2003–2008.

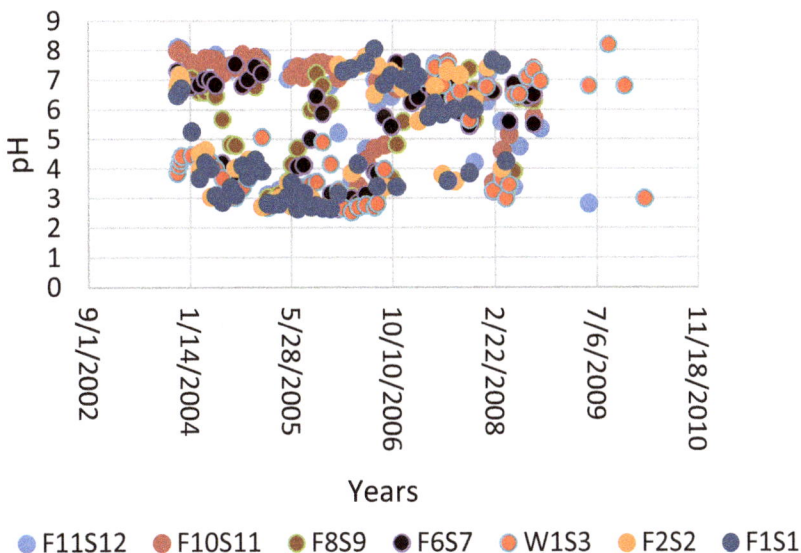

Figure 17. pH trending for all monitoring points.

The study area is critical in its strategic position because, apart from preserving its endowed environments, it is also because of its proximity to the Vaal Basin, which is the heartland of South Africa's economic activities as indicated in the digital elevation map (see **Figure 21**). Thus, measures and actions focusing on managing Tweelopiespruit pollution could also potentially benefit the Vaal System.

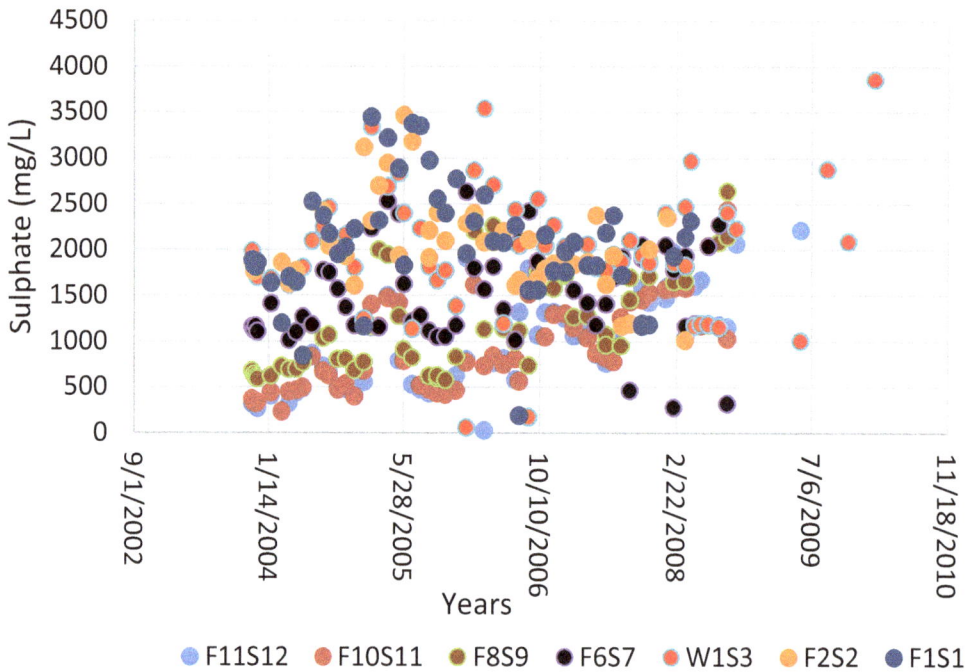

Figure 18. Sulfate trending for all monitoring points.

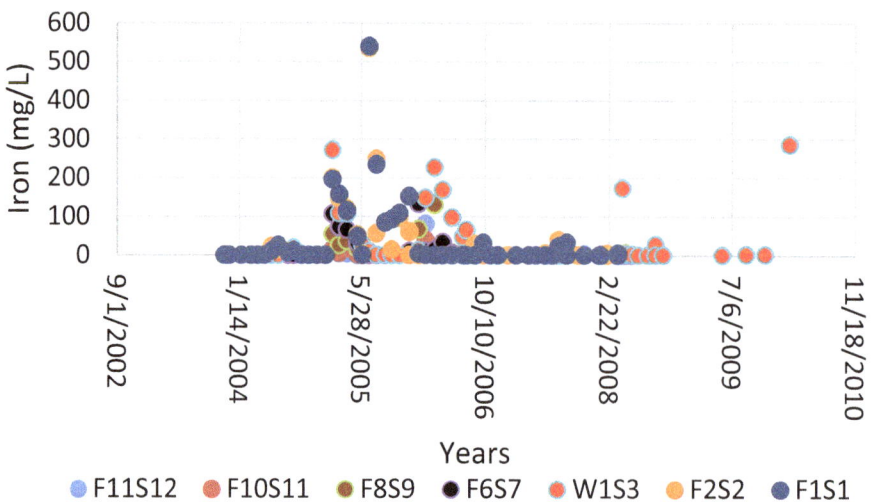

Figure 19. Iron trending for all monitoring points.

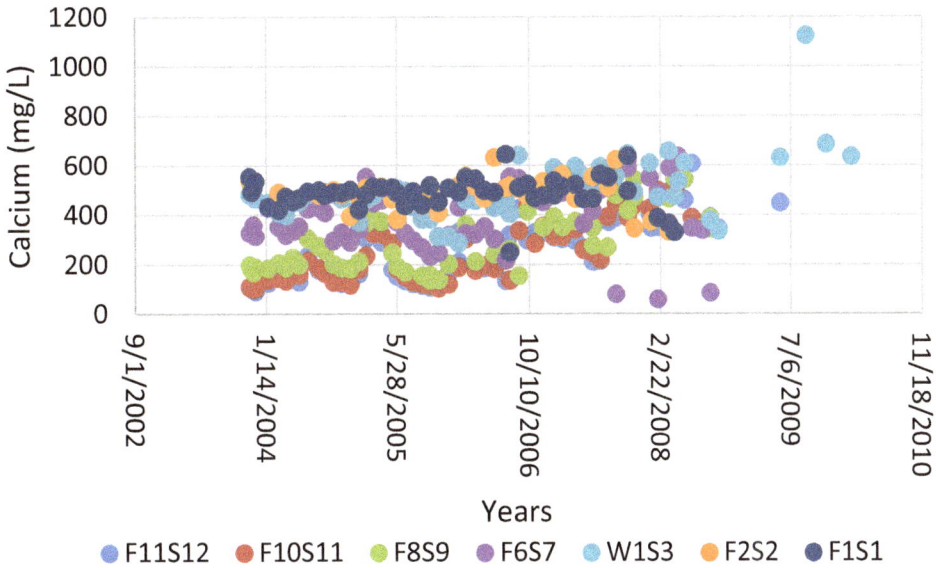

Figure 20. Calcium trending for all monitoring points.

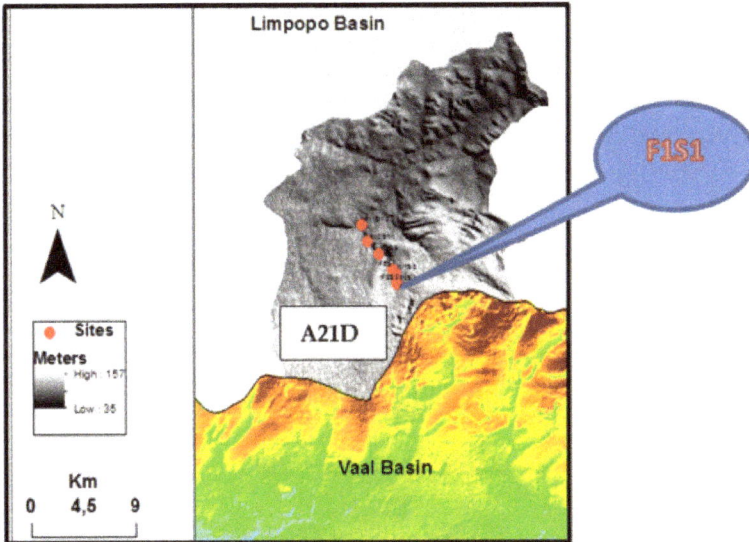

Figure 21. Overlaid monitoring points along Tweelopiespruit on micro-catchment A21D.

6. Conclusions

Pollution in the Krugersdorp Game Reserve is very significant as indicated by the chemical analysis' results for the monitoring points. Treatment of the polluted effluent does not seem to have an impact on the effluent for the study period up to 2008.

This research could benefit from the land use change detection, especially from satellite images which could show the devastating effects of the AMD on the environment within the Tweelopiespruit micro-catchment. These satellite images, freely available from the USGS website, could inform on the worsening situation in the micro-catchment.

It can be noted that peace-meal treatment works at the Randfontein site do not seem to have made a noticeable impact on pollution of the "dead" Tweelopiespruit, from AMD. The watershed, spanning the Witwatersrand System and the Upper reaches of the quaternary catchment A21D, is a hot spot for environmental disaster that is set to impact outer ecosystems for many years to come, and South Africa will have to pay for non-stringent environmental legislation, which was in place before the 1990s. The research results and conclusions aim to provide a baseline for critiquing ongoing research in the Tweelopiespruit micro-catchment in order to assist with answering the research questions that were initially raised against each objective. The use of satellite and remote sensing methods are recommended for further research.

Acknowledgements

The following people are most sincerely acknowledged for contributing to this paper in various but very valuable forms: Prof. Fred A.O. Otieno from Masinde Muliro University, Ms. Malebo D. Matlala, and Prof. Kevin Mearns – both from UNISA, Prof. Jannie Maree from Tshwane University of Technology, Mr. Phillip Hobbs, and the entire team at the Randfontein experimental treatment plant. The following organizations are acknowledged sincerely: Department of Water and Sanitation and Hydrology Section, USGS, CGIAR Consortium for Spatial Information, ESRI-South Africa, South African Weather Services and the South African Demarcation Board.

Author details

Bloodless Dzwairo[1]* and Munyaradzi Mujuru[2]

*Address all correspondence to: ig445578@gmail.com

1 Durban University of Technology, Department of Civil Engineering (Midlands), Durban, South Africa

2 University of Limpopo, Department of Water and Sanitation, Faculty of Science and Agriculture, Sovenga, Medunsa, Polokwane, South Africa

References

[1] Costello, C., Acid mine drainage: Innovative treatment technologies. National Network of Environmental Management Studies Fellow for US Environmental Protection Agency, Washington, DC, 2003.

[2] Tutu, H., T.S. McCarthy, and E. Cukrowska, The chemical characteristics of acid mine drainage with particular reference to sources, distribution and remediation: The Witwatersrand basin, South Africa as a case study. Applied Geochemistry, 2008. **23**(12): p. 3666–3684.

[3] Akcil, A. and S. Koldas, Acid Mine Drainage (AMD): causes, treatment and case studies. Journal of Cleaner Production, 2006. **14**(12–13): p. 1139–1145.

[4] Benner, S.G., W.D. Gould, and D.W. Blowes, Microbial populations associated with the generation and treatment of acid mine drainage. Chemical Geology, 2000. **169**(3–4): p. 435–448.

[5] Manders, P., L. Godfrey, and P. Hobbs, Acid Mine Drainage in South Africa, in Briefing Note 2009/02. 2009, CSIR Natural Resources and the Environment: Pretoria.

[6] Cole, M., et al., Development of a small-scale bioreactor method to monitor the molecular diversity and environmental impacts of bacterial biofilm communities from an acid mine drainage impacted creek. Journal of Microbiological Methods, 2011. **87**(1): p. 96–104.

[7] Benedetto, J.S., et al., Monitoring of sulfate-reducing bacteria in acid water from uranium mines. Minerals Engineering, 2005. **18**(13–14): p. 1341–1343.

[8] Hallberg, K.B., New perspectives in acid mine drainage microbiology. Hydrometallurgy, 2010. **104**(3–4): p. 448–453.

[9] McCarthy, T.S., The impact of acid mine drainage in South Africa. South African Journal of Science, 2011. **107**(5-6): p. 01–07.

[10] ESRI, Environmental Systems Research Institute (ESRI) ArcGIS for Desktop 10.2.

[11] Harris, J. and M. Andrew, The Top Ten of the Toxic Twenty, in The World's Worst Toxic Pollution Problems Report. 2011, Blacksmith Institute and Green Cross Switzerland: New York.

[12] Harris, J. and M. Andrew, The Top Ten of the Toxic Twenty, in The World's Worst Toxic Pollution Problems Report. 2011, Blacksmith Institute and Green Cross Switzerland: New York.

[13] Zilles, J.L., J. Peccia, and D.R. Noguera, Microbiology of Enhanced Biological Phosphorus Removal in Aerated-Anoxic Orbal Processes. Water Environment Research, 2002. **74**(5): p. 428-436.

[14] Wei, X., H. Wei, and R.C. Viadero, Post-reclamation water quality trend in a Mid-Appalachian watershed of abandoned mine lands. The Science of The Total Environment, 2011. **409**(5): p. 941–948.

[15] Kruse, N.A., et al., The Lasting Impacts of Offline Periods in Lime Dosed Streams: A Case Study in Raccoon Creek, Ohio. Mine Water and the Environment, 2012. **31**(4): p. 266–272.

[16] Donato, D.B., et al., A critical review of the effects of gold cyanide-bearing tailings solutions on wildlife. Environment International, 2007. **33**(7): p. 974–984.

[17] McCarthy, T., et al., Mine water management in the Witwatersrand gold fields with special emphasis on acid mine drainage. Report to the inter-ministerial committee on acid mine drainage. 2010: Pretoria. p. 1–128.

[18] Statutes of The Republic of South Africa - Constitutional Law: Act No. 108 of 1996. 18 December 1996: South Africa. p. 1241–1331.

[19] DEAT, No. 36 of 1998: National Water Act., in Government Gazette No. 19182, Vol. 398, August 1998. National Water Act, 1998. 1998, Department of Environmental Affairs and Tourism (DEAT): Cape Town.

[20] DEAT, No. 107 of 1998: National Environmental Management Act. Department of Environmental Affairs and Tourism, in Government Gazette No. 33306, Vol. 540, No. 9314 of 2010: National Environmental Management Act. 1998, Department of Environmental Affairs and Tourism (DEAT): Pretoria.

[21] DEAT, No. 53 of 2008: National Radioactive Waste Disposal Institute Act, in Government Gazette No. 31786, Vol. 523 number 19, January 2009. National Radioactive Waste Disposal Institute Act, 2008. 2008, Department of Environmental Affairs and Tourism (DEAT): Cape Town.

[22] DEAT, No. 28 of 2002: Mineral and Petroleum Resources Development Act, in Government Gazette No. 23922, Vol. 448, No. 1273 10 October 2002. Mineral and Petroleum Resources Development Act, 2002. 2002, Department of Environmental Affairs and Tourism (DEAT): Cape Town.

[23] Bologo, V., J. Maree, and F. Carlsson, Application of magnesium hydroxide and barium hydroxide for the removal of metals and sulphate from mine water. Water SA, 2012. 38(1): p. 23–28.

[24] De Beer, M., et al., Acid mine water reclamation using the ABC process. 2010, Council for Scientific and Industrial Research (CSIR): Pretoria.

[25] Motaung, S., et al., Recovery of Drinking Water and By-products from Gold Mine Effluents. Water Resources Development, 2008. 24(3): p. 433–450.

[26] Winde, F. and I. Jacobus van der Walt, The significance of groundwater-stream interactions and fluctuating stream chemistry on waterborne uranium contamination of streams–a case study from a gold mining site in South Africa. Journal of Hydrology, 2004. 287(1-4): p. 178–196.

[27] Dzwairo, B., Modelling raw water quality variability in order to predict cost of water treatment, in Department of Civil Engineering. 2011, Tshwane University of Technology: Pretoria. p. 237.

[28] Naicker, K., E. Cukrowska, and T.S. McCarthy, Acid mine drainage arising from gold mining activity in Johannesburg, South Africa and environs. Environmental Pollution, 2003. 122(1): p. 29–40.

[29] Liefferink, M., Greater intervention needed to tackle acid mine drainage, in Creamer Media's Mining Weekly. 2012.

[30] Kalin, M. and W.L. Caetano Chaves, Acid reduction using microbiology: treating AMD effluent emerging from an abandoned mine portal. Hydrometallurgy, 2003. 71(1–2): p. 217–225.

[31] Bologo, V., Treatment of acid mine drainage using magnesium hydroxide and barium hydroxide, in Department of Environmental, Water and Earth Sciences. 2012, Tshwane University of Technology: Pretoria.

[32] Baruah, B.P. and P. Khare, Mobility of trace and potentially harmful elements in the environment from high sulfur Indian coal mines. Applied Geochemistry, 2010. 25(11): p. 1621–1631.

[33] Martins, M., et al., Characterization and activity studies of highly heavy metal resistant sulphate-reducing bacteria to be used in acid mine drainage decontamination. Journal of Hazardous Materials, 2009. 166(2–3): p. 706–713.

[34] Benzaazoua, M., et al., The use of pastefill as a solidification and stabilization process for the control of acid mine drainage. Minerals Engineering, 2004. 17(2): p. 233–243.

[35] Luptakova, A. and M. Kusnierova, Bioremediation of acid mine drainage contaminated by SRB. Hydrometallurgy, 2005. 77(1–2): p. 97–102.

[36] Burford, M.A., et al., Correlations between watershed and reservoir characteristics, and algal blooms in subtropical reservoirs. Water Research, 2007. 41(18): p. 4105–4114.

[37] Hobbs, P. and J.E. Cobbing, The hydrogeology of the Krugersdorp Game Reserve area and implications for the management of mine water decant, in Groundwater Conference, 8-10 October 2007. 2007: Bloemfontein. p. 10.

[38] Davies, T.C. and H.R. Mundalamo, Environmental health impacts of dispersed mineralisation in South Africa. Journal of African Earth Sciences, 2010. 58(4): p. 652–666.

[39] Durand, J.F., The impact of gold mining on the Witwatersrand on the rivers and karst system of Gauteng and North West Province, South Africa. Journal of African Earth Sciences, 2012. 68(0): p. 24–43.

[40] Oberholster, P.J., et al., An ecotoxicological screening tool to prioritise acid mine drainage impacted streams for future restoration. Environmental Pollution, 2013. 176(0): p. 244–253.

[41] Khorasanipour, M., F. Moore, and R. Naseh, Lime treatment of mine drainage at the sarcheshmeh porphyry copper mine, Iran. Mine Water and the Environment, 2011. 30(3): p. 216–230.

5

The Behaviour of Natural and Artificial Radionuclides in a River System: The Yenisei River, Russia as a Case Study

Lydia Bondareva, Valerii Rakitskii and Ivan Tananaev

Additional information is available at the end of the chapter

Abstract

The Yenisei River is one of the largest rivers in the world. There is Mining and Chemical Combine (MCC) of Rosatom located at Krasnoyarsk, on the bank of the River Yenisei, 50 km downstream of the city of Krasnoyarsk. Since 1958 MCC used river's water for cooling of industrial nuclear reactors for the production of weapon plutonium—^{238}Pu. Besides the pollution caused by industry-related radionuclides, pollution by natural radionuclide—uranium and its isotopes— are also investigated. Besides the natural uranium isotopes (^{234}U, ^{235}U, ^{238}U), exclusive artificial isotope—^{236}U was also found. Yenisei water was also polluted by high tritium content: from 4 Bq/L (back road value) to 200 Bq/L (some sample of water). The total amount of radionuclides investigated was about 20 radioisotopes. These radionuclides have different physical and chemical properties, different half-lives, and so on. Thus, the data on artificial radionuclides entering the Yenisei River water were obtained by long-term monitoring, which is likely to be connected with the activity of the industrial enterprises located on the river's banks of the studied area.

Keywords: Yenisei River, migration, radionuclide, Siberia, isotopes, Russia

1. Introduction

The major part of population of Krasnoyarskii region lives on the banks of Yenisei River. Yenisei is—one of the largest rivers in the World: its length from junction of Big Yenisei and Small Yenisei is 3487 km, from Small Yenisei's rise—4287 km and from Big Yenisei's rise—4123 km. The place of junction of Big and Small Yenisei near city of Kyzyl is considered as geographical centre of Asia. Rising in the south, in the mountain deserts of Mongolia, Yenisei flows in the north direction for nearly 3000 km, crosses various latitudinal geographical zones, falls into the Arctic Ocean, forming estuary zone up to 30 km wide. Length of Yenisei exceeds the same of Danube River (2857 km), Mississippi (3770 km) and Indus (3180 km). Yenisei River is the most affluent river of Russia with a runoff rate of 624 km^3/year. Mean water consumption in the estuary is 19,800 m^3/s and the maximal value is 190,000 m^3/s. With respect to basin area (2580 thousand km^2) Yenisei holds second place (after the Ob) and the seventh place among all rivers of the world. The nominal border between Western and Eastern Siberia lies along Yenisei. There are three hydroelectric power plants (HPP) on the Yenisei River and on the rivers falling into it. River's waters are characterized by high transparency (up to 3 m) and low mineralization (mean value is 54 mg/l) and also by high oxygen concentration. Flow velocity and river width can change considerably: from 1.5 to 12–15 km/h and from 0.2–0.5 to 3–5 km, respectively. Solids of the channel in the uppers are faceted soils that are changed into gravelly sand in the middle course and into sandy-clay in the lower course near the fall into the Arctic Ocean.

There is a constant mixing of water layers because of hydroelectric power plant's activity, thus not affecting water temperature from the depth of water flow even on higher distances after HPP stanch. At the beginning of July, water temperature in Krasnoyarsk district and after 100–150 km further down the course is ~10°C, at the end of July–August it is 15–17°C. River's ecosystem is related to oligotrophy with fauna-rich river, there are more than 500 species of algae and diatoms [1].

There is Mining and Chemical Combine (MCC) in Rosatom, located at Krasnoyarsk, on the bank of the River Yenisei in 50 km downstream of the city of Krasnoyarsk. There are atomic reactors and radiochemical production in the MCC. Since 1958 MCC used water for cooling industrial nuclear reactors for the production of weapon plutonium –^{238}Pu. River water, while passing through the cooling system of reactors, returned to Yenisei. Effluent waters contained a great amount of radionuclides that were formed during neutron activation of traces (solid slurry and dissolved compounds), which are present in river water. Two direct flow reactors were withdrawn in 1992, because the activity level of the effluent waters of MCC was remarkably decreased.

2. Radionuclides (natural and artificial) in the streams of the Yenisei River

As a result of long-term activity of MCC, the Yenisei's ecosystem contains considerable amounts of industry-related radionuclides [2]. In particular, an increased level of radioisotope contents in bed deposits and alluvial soils was found [2–6] and distribution and migration of radionuclides both in near-field influence of MCC [7, 8] and in significant distance away from effluent zone, including estuary of Yenisei, were indicated. As early as in the beginning 1970, the pollution zone of Yenisei's bottom land by ^{137}Cs was found by airborne gamma survey. In district of Yeniseysk city (island Gorodskoi around 300 km downstream of MCC), the specific activity of ^{137}Cs reaches 16,300 Bk/kg in some places, power of exposure (PE)−270 µR/h. According to present standards, bottom sediments and alluvial soils at this region are related to solid radioactive wastes. ^{137}Cs is the main radionuclide polluting soils and bottom sediments are $^{152+154}$Eu and ^{60}Co [9].

In this chapter, the results of research conducted mainly in the middle course of Yenisei in the 15 km region (from fall place of Ploskiy river (0 km) to Bolshoy Balchug (15 km), **Figure 1**) are described. In this region, at a water flow rate $Q = 4085$ m^3/s the depth and current velocity were defined as $H \approx 7$ m, $v = 1.25$–1.8 m/s, respectively. Jet with industrial wastes spends along the right bank not more than 0.1 of river's width, i.e. along bottom land, where current velocity and depth are several times lower.

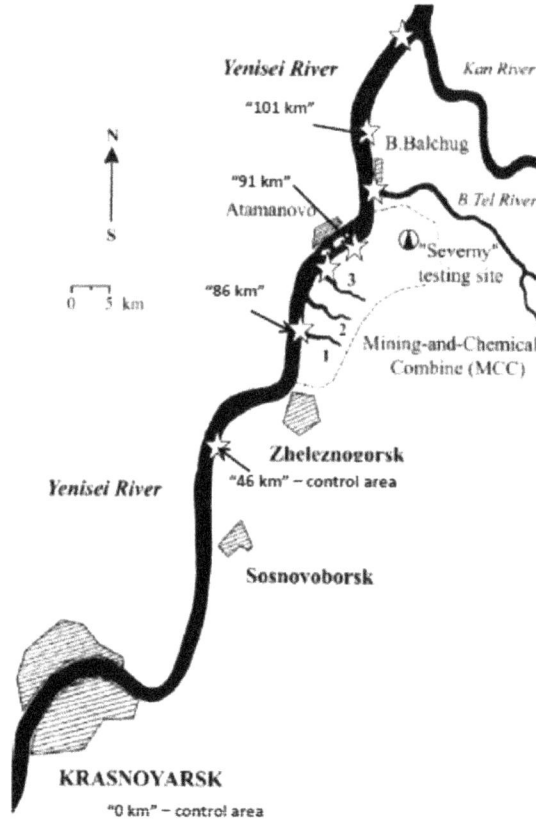

Figure 1. Sketch-map of the some region of the Krasnoyarsk Territory near the Mining—Chemical Combine of the Rosatom—surface water of the Yenisei River basin. 1: Shumikha River; 2: Stream No. 2; 3: the Ploskii Stream; – – – – : the boundary of the MCC sanitary-protective zone. ☆point of collection. Sampling points: '0 km'—56°27′05″N, 93°36′31″E; '2 km'—56°23′18″N, 93°37′13″E, '5 km'—56°23′40″, '15 km'—56°27′05″, 93°42′22″E.

2.1. Uranium: natural and artificial

Besides the pollution caused by industry-related radionuclides, pollution by natural radionuclide—uranium and its isotopes are also investigated.

The total uranium content is the main factor to determine the radiation level of water sources, its value is standardized and controlled by ecological services. Uranium in water is truly dissolved and found in the form of uranyl carbonate complex anions. In general, river waters contain 600 ng/l of dissolved uranium. Despite that main natural transport agents—water carries uranium in small amounts, one should not exclude that there can be local transfers of uranium in significant amounts [10].

The main feeders, contributing to the radioactive pollution of the Yenisei, are majorly the right bank feeders, situated near MCC outlet: river Kan, on the bank of which the electrochemical plant (ECP, Zelenogorsk city) is situated, and river Bolshaya Tel', flowing along the border of testing area 'Sverniy' MCC (Zheleznogorsk city).

According to data, presented in the monograph [9], the most of the region's waters, related to the bottomland of Kan, contain from 0.04 to 3 µg/l of uranium that is considered as highly pure with respect to natural radionuclide content. In addition, there was no trend in uranium content from the location of selection. Only in one place at the turn of Kan's course to the north vs. course of Bogunay river, it was revealed that all of the waters contain uranium from 1 to 3.3 µg/l. Industrial waters discharged by ECP into Kan near the plant administration were similar to natural uranium content and contained 0.05–0.08 µg/l of uranium.

Natural stream feeding Syrgyl river contained from 0.03–0.07 to 1.0–7.3 µg/l of uranium. The contents in the range 0.3–5.0 µg/l were shown to be natural geochemical background of uranium in the studied region, in particular, in the bottomland of Kan. All of the excesses are considered as abnormal.

The analysis data [1–12] shows that the geochemical background level of uranium in the Yenisei River is in agreement with the mean statistical level for the basins with major contribution of natural uranium resources, e.g. Baikal Lake and rivers of Altai region: from 0.15 to less than 2.0 µg/l.

Uranium content in waters which were collected from Bol'shaya Tel' in the September 2007 at the 1000 m place from the estuary is 3–60 times higher than values obtained for uranium (mean value 0.33 ± 0.08 µg/l) in background samples (Yenisei, tideway). Moreover, this period was indicated by significantly higher uranium concentrations as compared with other studied months. This increase becomes remarkable for the 1000 m place, where uranium concentration is 16 µg/l that is very close to the accepted in Canada and Australia standards for the minimal allowed uranium concentration—20 µg/l and by 8 times exceeds accepted by WHO standard—2 µg/l. Despite that obtained values are lower than the level of exposure (LE = 75 µg/l) accepted by in NRS of Russian Federation [9, 10], uranium concentration in some places of Bol'shaya Tel' in September is, in general, can be considered as abnormal. It is known that natural uranium is a mixture of three isotopes: ^{238}U—99.2739% ($T_{1/2} = 4.468 \times 10^9$ years), ^{235}U—0.7024% ($T_{1/2} = 7.038 \times 10^8$ years) and ^{234}U—0.0057% ($T_{1/2} = 2.455 \times 10^5$ years). In contrast to other isotope pairs, last two isotopes are in constant proportion, regardless of high migration activity of uranium and geography: $^{238}U/^{235}U = 137.88$ [13, 14]. The presence of uranium was truly established in the waters of Bol'shaya Tel', it can only be originated artificially: in the sample from 1000 m (October 2006) ~0.05 ng/l and in the sample from Bol'shaya Tel' (March 2007) ~0.03 ng/l. In addition, the ratio of $^{236}U/^{234}U$ at these places is 1:0.8, respectively.

Besides, water samples obtained in September provided information about anion content of NO_3^- (~2 mg/l, while the maximum permissible concentration (MPC) is 45 mg/l), CH_3COO^- (~7 mg/l) in the waters of Bol'shaya Tel' (1000 m from estuary). It is considered that the presence of such anions can indicate the non-equilibrium conditions in basin solution. Such situation is considered rather usual for liquid radioactive wastes, where acetate and nitrate, due to kinetic limitations of the acetate oxidation by nitrate, can coexist even at high (about 100°C) temperatures [10].

Generalized information about the total uranium content in water samples of the Yenisei River is given in **Figure 2**.

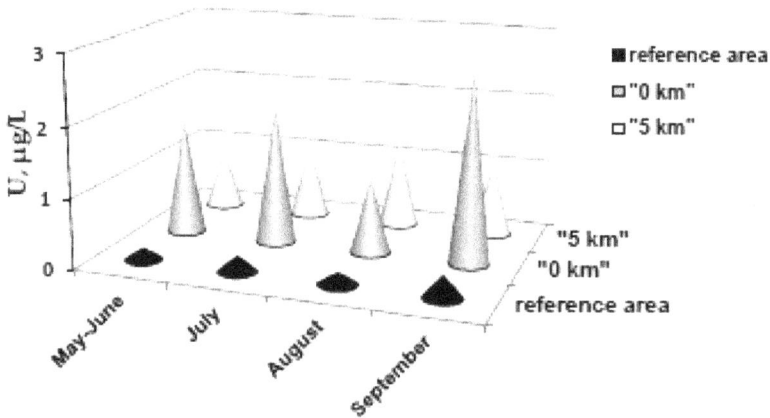

Figure 2. Results of determination of total uranium content in Yenisei water at distances from water discharge of MCC '0 km' and '5 km', taken 2006–2009, 'distance from water discharge MCC'.

Presented data indicate uranium content in the estuary of Ploskiy river '0 km' to exceed by 6–9 times background values of uranium typical for Yenisei. Further investigation of isotope composition of indicated water samples revealed that a ratio of uranium isotopes differ from natural isotopes and also the presence of ^{236}U can also evidence the industrial origin of high uranium concentrations as compared with background values. Isotope analysis of some samples has been carried out.

In water samples of Yenisei (pick point '0 km') the ratio of $^{238}U/^{235}U$ is 119:120. Besides, artificial uranium isotope ^{236}U ($T_{1/2} = 2.39 \times 10^7$ years) was found, the ratio of which to ^{234}U equals $^{236}U/^{234}U$ ~0.1–0.2. Thus, one can state that high uranium concentration in Yenisei waters is caused by MCC activity.

2.2. Tritium and other radionuclides

2.2.1. Tritium

Besides artificial radionuclides, Yenisei water was also polluted high tritium content. To prove this, the tritium content was determined in the picked water samples. Results are given in **Figure 3**.

Tritium content in the picking site '0 km' exceeds by 15–20 times the background tritium content obtained via long-term monitoring and typical for Yenisei (4 ± 2 Bk/l) [15–20].

To prove industrial origin of tritium in water samples it is recommended to control content of gamma-emitting radionuclides. There is significant amount of artificial radionuclides in the studied water.

2.2.2. Radionuclides without tritium

Depending on the state of radionuclides that can be present as simple ions to molecules and hydrolyzed forms, colloids and pseudocoolloids, organic and inorganic particles [21, 22] and, respectively, migrates over long distances and be sorbed by ecosystem immediately near the discharge area. Content of TUE in surface basins is extremely low and equals 10^{-10}–10^{-15} M, within limits of the most sensitive spectral techniques, e.g. mass-spectrometry [23, 24]. For the precise determination of TUE contents as well others radionuclides such as ^{90}Sr in water systems, the most frequently used methods are hybrid ones, combining preliminary concentrating and separating of radioisotopes with various detecting methods, e.g. alpha-, beta- and gamma-spectrometry [25–27].

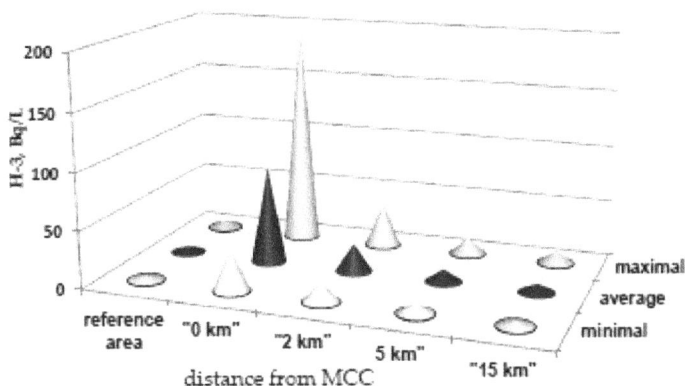

Figure 3. Average tritium content in water samples of Yenisei (distance down the stream from places of water discharge by MCC).

To increase the number of identified radionuclides, the method for concentrating the radionuclide from Yenisei water samples has been introduced [8]. Data obtained after concentration of water samples is given in **Tables 1** and **2**.

Water samples contain the bunch of artificial radionuclides. To increase the number of identified radionuclides, the method for concentrating the radionuclide from Yenisei water samples has been improved [8].

The method for concentrating the radionuclide was accepted on the basis of two widely known methods of co-precipitation with oxyhydroxide of Fe (III) and Mn (IV) oxide [28, 29].

Artificial radionuclides, which have different origin, have been found in water samples: induced (activated) radionuclides—^{24}Na, ^{46}Sc, ^{51}Cr, ^{54}Mn, ^{59}Fe, ^{60}Co, ^{65}Zn, ^{76}As and others; satellite radionuclides—^{99}Mo, ^{124}Sb, ^{131}I, ^{133}I, ^{141}Ce, ^{144}Ce and others. The most distinctive are trans-uranic radionuclides—^{239}Np, isotopes of Pu. In water samples, taken down the stream from MCC (5 km), besides the decreasing concentration of artificial radionuclides there were found some natural radionuclides: ^{210}Pb and ^{232}Th. There were included the presence of long-living satellite isotope ^{152}Eu ($T_{1/2}$ = 13.6 years) ~ 0.04–0.06 Bk/l and the presence of short-living activated radionuclide ^{58}Co ($T_{1/2}$ = 71.3 days) ~ 0.03–0.07 Bk/l in water samples.

2.3. Suspended matter of the Yenisei River: trucks for transport of radionuclides in the water flow

Because major part of radionuclides has been found in the suspended matter, transporting by water stream of Yenisei, more thorough studies of suspended matter of Yenisei have been conducted.

№	Isotopes	2007	2008	2009
1	^{24}Na	2.5 ± 1.4	1.9 ± 0.2	–
2	^{46}Sc	0.21 ± 0.01	0.136 ± 0.006	0.086 ± 0.06
3	^{51}Cr	6.0 ± 0.2	2.7 ± 0.1	3.4 ± 0.1
4	^{54}Mn	0.014 ± 0.003	0.014 ± 0.003	0.007 ± 0.002
5	^{59}Fe	0.16 ± 0.01	0.11 ± 0.008	0.07 ± 0.01
6	^{60}Co	0.13 ± 0.01	0.17 ± 0.008	0.09 ± 0.01
7	^{65}Zn	0.11 ± 0.01	0.055 ± 0.007	0.03 ± 0.004
8	^{76}As	8.5 ± 0.6	4.5 ± 0.2	4.7 ± 0.6
9	^{85}Sr	–	0.014 ± 0.003	0.003 ± 0.001
10	^{99}Mo	–	0.093 ± 0.008	0.04 ± 0.01
11	^{103}Ru	0.027 ± 0.004	0.026 ± 0.003	0.012 ± 0.006
12	^{106}Ru	0.078 ± 0.025	–	0.04 ± 0.01
13	^{124}Sb	0.016 ± 0.003	0.020 ± 0.003	0.012 ± 0.004
14	^{131}I	0.051 ± 0.013	0.031 ± 0.005	0.028 ± 0.008
15	^{133}I	–	–	0.14 ± 0.02
16	^{137}Cs	0.057 ± 0.005	0.142 ± 0.009	0.09 ± 0.02
17	^{141}Ce	0.048 ± 0.006	0.050 ± 0.006	0.021 ± 0.007
18	^{144}Ce	0.08 ± 0.02	0.13 ± 0.02	0.04 ± 0.01
19	^{239}Np	29.5 ± 1.4	17.1 ± 0.3	10.3 ± 0.8

Table 1. Radionuclide content in water samples after concentrating, taken in the place of MCC discharge ("0 km"), Bk/l.

N	Isotopes	2006	2007	2008	2009
1	^{24}Na	–	–	–	0.07 ± 0.025
2	^{46}Sc	0.11 ± 0.02	0.09 ± 0.02	–	0.002 ± 0.001
3	^{51}Cr	2.6 ± 0.2	1.4 ± 0.2	0.037 ± 0.013	0.057 ± 0.009
4	^{58}Co	–	–	–	–
5	^{60}Co	0.14 ± 0.02	0.11 ± 0.04	0.006 ± 0.001	0.002 ± 0.001
6	^{65}Zn	0.10 ± 0.03	0.07 ± 0.02	0.003 ± 0.001	0.005 ± 0.002
7	^{76}As	3.1 ± 0.3	0.08 ± 0.03	1.07 ± 0.08	0.103 ± 0.015
8	^{106}Ru	0.3 ± 0.1	0.4 ± 0.3	–	0.0064 ± 0.0061
9	^{131}I	0.04 ± 0.01	–	0.002 ± 0.001	0.0023 ± 0.0009
10	^{137}Cs	0.07 ± 0.02	0.04 ± 0.01	0.001 ± 0.001	0.0015 ± 0.0013
11	^{140}La	0.16 ± 0.03	0.08 ± 0.03	–	0.006 ± 0.002
12	^{144}Ce	0.25 ± 0.07	0.04 ± 0.02	–	–
13	^{152}Eu	0.06 ± 0.02	0.04 ± 0.02	–	–
14	^{239}Np	0.27 ± 0.02	0.32 ± 0.04	0.39 ± 0.02	0.261 ± 0.007

Table 2. Radionuclide content in water samples, taken from Atamanovo region, after concentrating (taken at 5 km down the stream from the place of discharge), Bk/l.

The investigations were carried out in the middle reach of the River Yenisei at the site 15 km (from the inflow of the Plosky stream (0 km) to the village Bolshoy Balchug (15 km) (**Figure 1**). The stream with technogenic admixtures propagates along the bank of the river not more than 0.1 one-tenth of the width of the river, i.e. along the flood plain where the river flow speed and the width are several times less.

As a result of ultra-filtration method, it was found that the main part of the suspended particles (up to 90%) was concentrated in the pelitic fraction of >5 μm. The filters with the suspensions were fixed on the specimen mount with the help of the conducting double-sided adhesive carbon type and placed into the electron microscope chamber. The precipitate was found to contain particles of quartz, mica and iron-containing minerals (limonitic and magnetic iron), mainly, with the size not exceeding 10–15 μm. Moreover, the precipitate revealed the presence of a considerable amount of various biological objects (diatoms, annelids, plant spores, etc.). All the mineral particles and biota were covered with a layer of fine limonitic-clayish particles. Spectral analysis of some parts of the sample (selected particles, characteristic details) was carried out. The suspended matter contains a large colony of diatoms, for example, *Meridion circulare, some cyclotellas* and *opyphoros, Cyclotella vor. Jacutca* (**Figure 4**).

The fraction with the size of '5-1 μm' uniformly covers the filter surface with a layer of fine particles. The precipitate mainly consists of mineral components (calcite, clays, clayish minerals, quartz and gypsum debris).

The fraction '1.0–0.2 μm' uniformly covers the filter surface with a layer of fine particles of the micron and submicron size, they are mainly aluminosilicate compounds having various structure and composition, limonite, calcite and gypsum.

Figure 4. Material composition of the water suspensions (separated by the ultra-filtration method). The fraction ≥5 μm. Magnification power of 2000×.

The material composition of the solid suspensions in the Yenisei River water generally corresponds to the mineral compositions of the rocks and the products of their hypergenesis which collected from the channel and the banks of the river. Occasionally, the admixture of the particles of technogenic origin (ash wastes from boiler stations) is observed.

Thus, it was shown that the suspended substance is similar to its geomorphology with the bottom sediments of the Yenisei River. However, the suspensions entering the river with the industrial discharge water significantly differ from the suspensions of the mainstream both in their composition and particle size.

At the sampling of the district runoff of radionuclides when the time of the discharge contact with the river water was insignificant, the radionuclides ^{3}H, ^{24}Na, ^{60}Co, ^{239}Np and ^{99}Mo (~90%) were mainly presented as a fraction <0.2 μm (filtrate). These can be both free ions in the molecular solution (e.g., ^{24}Na$^+$), and molecules or sorbed ions in colloid particles which managed to pass through a 0.2 μm filter. ^{46}Sc, ^{214}Bi, ^{103}Ru are mainly presented in solid phase, while the last two isotopes being in the coarsest fraction (more than 90% of them). ^{85}Sr and ^{131}I have less uniform phase distribution. ^{76}As is almost absent in the most coarse fraction (>5 μm). In the samples taken 5 km downstream, there is a decrease of the total activity, first of all, due to the coarse particle sedimentation. The radionuclide redistribution according to the size fractions was found: almost the whole amount of ^{60}Co is concentrated in the fraction with the size of >1 μm, a considerable amount of ^{214}Bi is transformed into a solution (the fraction <0.2 μm), almost 40% of ^{99}Mo and up to 70% of ^{24}Na are transformed into the fraction of 1–0.2 μm. With the total background level decrease there appear natural radionuclides ^{212}Pb and ^{234}Th in the solid phase as well as ^{65}Zn in the solution.

3. Mathematical calculations of the mass transport of technogenic radionuclides in the water flow of the River Yenisei in the impact zone of the Mining and Chemical Combine

In the chapter, the results radiation-chemical situation in the middle reach of the Yenisei River located in the nearest zone of the influence of the Mining and Chemical Combine of Rosatom have been described. It has been shown that a wide range of radionuclides, heavy metals and organic substances of different genesis flow into the waters of the Yenisei River. It has been demonstrated that radionuclides and other pollutants are transported by the water flow in the form of molecular solution or colloids or with suspended matter. In this case, the suspended matter consists of pelitic finely dispersed mineral particles, plant and organic detritus and amounts of living biological objects.

Calculations have been made according to the described method in the area of the River Yenisei from the estuary of the river Plosky up to the island Atamanovsky. Assuming the water discharge to be $Q = 4085$ m^3/s the river depth $H \approx 7$ m and the flow rate $v = 1.25-1.8$ m/s in the given section are estimated based on the hydraulic model. According to an earlier estimation, the stream with the technogenic admixtures propagates along the right bank, not far than one tenth of the river width, i.e. along the flood plain where the flow rate and depth are several times lower than those calculated based on the hydraulic model. According to the calculations: $H_n \approx 2.5$ m, $v_n \approx 0.38$–0.44 m/s.

Transport of radionuclide along the Yenisei River is based on a modified one-dimensional model proposed by Schnoor et al. [30]. For the whole length of the Yenisei, a homogeneous distribution of radionuclides over the cross-section is presupposed. It is assumed that both in the water column and in the active sediment layer the radionuclides are present in two forms: soluble and adsorbed forms. The most important processes influencing the behaviour of radionuclides include adsorption and desorption, sedimentation of suspended particles from the river water and resuspension from the active sediment layer, activity exchange between the pore water of the sediment and overlying water due to diffusion through the boundary and radioactive decay.

The calculations presented in this chapter are limited to the abiotic form of substance transport since the contribution of the biogenic component is considered to be insignificant [9].

Complex fresh water systems, such as large rivers, are assumed to be composed of a chain of interconnected 'elementary segments (ES)' that are comprised of: (a) the water column, (b) an upper sediment layer strongly interacting with water ('interface layer'), (c) an intermediate sediment layer below the 'interface layer' ('bottom sediment'), (d) a sink sediment layer below the 'bottom sediment', (e) the right and left sub-catchments of each ES.

Depending on the water discharge rate and geometry of the river bed the stream velocity varies which determines the transport of the sediment suspensions and sediment disturbance-sedimentation. To estimate the accumulation of radionuclides in the bottom sediments, a mathematical model described by Belolipetsky and Genova was used [31].

The concentrations of radionuclides on solid particles were assumed to be proportional to the area of the particle surface. We used the field data the fraction distribution of radionuclides in the initial solution. Then, the particle transport and sedimentation along the river bed was estimated. In the channels and floodplain (in the areas with small stream velocities) there occurs sedimentation of the sediment suspensions. During the periods of the increased water discharge rate (spring floods, increased volume of the hydroelectric station), the sediment disturbance is also possible as well as transport of impurities downstream (secondary pollution).

To describe the sediment suspension transport in a turbulent flow of non-compressible liquid a simplified equation is used:

$$\partial Si/\partial t + u_в\, \partial Si/\partial x = qSi/h + q/\omega \cdot Siq \tag{1}$$

where S_i is the concentration of the ith fraction [kq/m³]; S_{iq} is the concentration of an impurity of the ith fraction, entering with the tributary on the way q; q_{si} is sediment disturbance-sedimentation of the impurity of the i-th fraction; t is time; x is a coordinate directed along the current; Q is the discharge rate; ω is the cross-section area of the river bed; $u_в = Q/\omega$ is the cross-section; average velocity h is the depth.

The bottom exchange is determined by the formula

$$qSi = (Si\, tr - Si0) \cdot wgi, \; Si\, tr = 0.01 \cdot \alpha i \cdot Str, \; qS = \sum qSj \tag{2}$$

$$S_{tr} = \begin{cases} 0.2 \cdot u_s/gh\, w_g, \text{ if } w_g < w_* \\ , \quad w_g = (\rho_s - \rho_s)/\rho_s \cdot g/18v \cdot d^2_{\ cp} \\ 0, \text{ if } w_g \geq w_* \end{cases} \tag{3}$$

The transport capability of the flow S_{tr} depends on the depth-average flow velocity, depth and hydraulic coarseness; q_s is the mass exchange with the bottom.; s_{i0} is the concentration of the i-th fraction near the bottom; α_i is the percent content of the i-th fractions in the bottom sediments. When calculating $S_{i\,tr}$ using Eq. (3) it should be taken into account that $S_{i\,tr}$ cannot exceed the concentration of the i-th fraction in the bottom sediments ($S_{i\,day}$), therefore, when $S_{i\,tr} > S_{i\,day}$ it is assumed that $S_{i\,tr} = S_{i\,day}$. If the concentration of the i-th fraction in the bottom sediments is equal to zero, then $S_{i\,tr} = 0$.

The main change in the bottom sediment composition is assumed to be due to sediment disturbance and sedimentation. When $q_s > 0$, the bottom sediments enter the flow (washing out, sediment disturbance) and when $q_s < 0$ the silting of the river bed is observed (sedimentation of the suspended particles).

Let z_* be the thickness of the active layer of the bottom sediments. Assuming that the formation of the upper layer of the bottom sediments (with the thickness z_*) results in the sediment disturbance-sedimentation, the mass conservation equation for the i-th fraction in the bottom sediments $S_{i\,day}$ is written as follows:

$$\partial(z_* \cdot S_{i\,day})/\partial t = -q_{Si} \tag{4}$$

Since $\sum q_{Si} = q_{S'} \sum S_{i\,day} = \rho$, from Eq. (4) one obtains the equation to find z_*:

$$\partial z_*/\partial t = -q_S/\rho \tag{5}$$

The calculation algorithm for the suspended and bottom sediment dynamics consists of the following stages:

Stage 1. The water flow rates u_w are determined as well as the depth h from the solution of the Saint-Venant equation.

Stage 2. Determination of the initial conditions. The granulometric composition of the bottom sediments in the section $X = X_j$ is taken to be $(d_i, a^0_{iday,j})$, where d_i is the diameter of the ith fraction particle (mm), $a^0_{iday,j}$ is the percentage of the i-th fraction in the bottom sediments, $i = 1, 2, ..., n$.

Stage 3. Establishment of the boundary condition in the initial section ($X = X_0$). In the initial section, $S^n_{iday,0}$ are determined using relations employed for the second stage, $S^n_{i,0}$ are estimated using the field data.

Stage 4. Estimation of the mass exchange between the bottom water and water flows. From the condition $w_{gi} \leq w_*$, $w_* = 0.4u_*$ one determines the fractions which are suspended. Let the suspended fractions be assigned the following index $i = 1, 2, ..., i_*$, $a_{i,j}$ is the percentage of the suspended fractions in the section. The percentage of all the suspended fractions is $r_j = a_{1,day,j} + a_{2,day,j} + ... + a_{i,day,j}$. Then, the percentage of the suspended ith fraction is

$$a_{i,j} = 100 \cdot r_j^{-1} \cdot a_{i,day,j}, i = 1, 2, ...i \tag{6}$$

If $r_j = 0$ (the suspended fractions are absent), then, all $a_{i,j} = 0$.

Stage 5. Estimation of the concentrations of the suspended and bottom sediments as well as the location of the water-bottom interface.

Stage 6. Calculation of the granulometric composition of the bottom sediment:

$$a_{i, \partial H, j} = S^{n+1} \cdot \rho^{-1} \cdot 100 \tag{7}$$

Stage 7. Estimation of the bottom sediment radioactive contamination in the calculation sections.

Each fraction is assumed to be uniformly contaminated by radionuclides:

$$R^n_{i,j} = \lambda_i S^n_{i,j} \tag{8}$$

Knowing the contamination level in the initial section $R^n_{i,0} = \lambda_i S^n_{i,0}$, it is possible to estimate the level of the radionuclide contamination in the sections downstream the river

$$R^n_{i,j} = S^n_{i,j} \cdot (S^o_{i,j})^{-1} \cdot R^n_{i,0} \tag{9}$$

In the next time interval, the calculations are repeated (from stage 3 to stage 7).

The influence of the suspension-sedimentation processes on the admixture transport in the river flow close to the right bank of the River Yenisei in the studied area has been estimated.

The calculations made show that the concentrations of the lightest fraction in the calculation area almost do not change, while for the heavier fractions the decline of the suspended sediment concentrations is observed and the level of the radionuclide contamination also decreases (**Table 3**).

In the field data, the increase of the coarse fraction concentration is observed which is not connected with the suspension-sedimentation process. (S_nat, R_nat are the measured values, S_calc, R_calc are the calculated ones)

d, мм	0.00020	0.00045	0.005	0.01
District reset MCC				
S^0, г/л	0.0001	0.0005	0.0043	0.0031
R^0, Бк/кг	118.904	0.1728	0.1165	1.8224
Island Atamanovsky				
S_nat	0	0.0001	0.0009	0.0583
S_calc	0.0001	0.0005	0.0039	0.0021
Island Atamanovsky				
R_nat	2.4727	0.01101	0.01596	0.0853
R_calc	118.9009	0.1727	01044	1.2065

Table 3. Concentrations of particulate matter size fractions: real and calculated data.

Thus, the abiogenic mass-transport of the technogenic radionuclides, metals being among them, occurs mainly due to the coagulation of the suspended particles and contamination redistribution into bigger fragments.

Our calculations show that the concentration of the lightest fraction of the water on the current site remains virtually unchanged. However, we observed that concentrations of suspended sediment had decreased for heavier fractions and, consequently, decreased the level of contamination. In addition, our field data indicated an increase in the concentration of coarse fraction, which is associated not only with the resuspension-deposition, but also with the coagulation of suspended solids.

Thus, the data on artificial radionuclides entering the Yenisei River water obtained by long-term monitoring, which is likely to be connected with the activity of the industrial enterprises located on the river's banks of the studied area.

Acknowledgements

This investigation was made with financially supported by the Russian Bureau of fundamental researches N-16-05-00205

Author details

Lydia Bondareva[1]*, Valerii Rakitskii[1] and Ivan Tananaev[2]

*Address all correspondence to: lydiabondareva@gmail.com

1 Federal Scientific Center of Hygiene named after F.F. Erismana, Moscow, Russia

2 Far Eastern Federal University, Vlodivostok, Russia

References

[1] Sposito G. The Chemistry of Soils. Oxford: Oxford University Press; 1989. 347 p.

[2] Bolsunovsky A., Bondareva L. Actinides and other radionuclides in sediments and sub-merged plants of the Yenisei River. Journal of Alloys and Compounds, 2007; **444–445**: 495–499.

[3] Vakulovsky S.M. Radioactive contamination of surface water in the territory of Russia in 1961–2008. In.: Problems of hydrometeorology and environmental monitoring. Ed. Vakulovsky S.M., Obninsk, Roshydromet. Russia; 2010; 2: 115–127.

[4] Kuznetsov Y.V., Revenko Y.A., Legin V.K. et al. By the estimation of the Yenisey River contribution to the total radioactive pollution of the Kara Sea. Radiochemistry, 1994; **36**: 546–553.

[5] Kuznetsov Y.V., Legin V.K., Shishlov A.A. et al. Investigation of behaviour of ^{239,240}Pu and ^{137}Cs in system Yenisei River - Kara Sea. Radiochemistry, 1999; **41**: 181–186.

[6] Kuznetsov Y.V., Legin V.K., Strukov V.N. et al. Transuranic elements in the floodplain sediments of the Yenisei River. Radiochemistry, 2000; **42**: 470–477.

[7] Bondareva L.G., Bolsunovsky A.Y. The study of modes of occurrence of man-made radionuclides ^{60}Co, ^{137}Cs, ^{152}Eu, ^{241}Am in the sediments of the Yenisey River. Radiochemistry, 2008; **50:** 475–479

[8] Bondareva L.G., Bolsunovsky A.Y., Trapeznikov A.V., Degermedzhy A.G. Using the new technique concentrating of transuranic elements in the Yenisei river water samples. Doklady Chemistry, 2008; **423**: 479–482.

[9] Sukhorukov F.V., Degermedzhy A.G., Belolypetsky V.M. et al. Patterns of distribution and migration of radionuclides in the valley of the Yenisei River. - Novosibirsk: SD RAS, GEO, 2004. 286 p. (in Russian)

[10] Broder J. Merkel, Britta Planer-Friedrich, Christian Wolkersdorfer (ed). Uranium mining and hydrogeology III, International Mine Water Association Symposium, 15–21 September 2002, Freiberg/Germany (Preventing uncontrolled spread of radionuclides into the environment, Novosibirsk: SD RAS NIC OIGGM, 1996).

[11] Fedorin M.A. Multiwave XRF-SR determination of U and Th in bottom sediments of Lake Baikal: Brunhes paleoclimatic chronology. GEOL GEOFIZ, 2001; **42**(1–2): 186–193 (in Russian)

[12] Egorova I.A., Puzanov A.V., Blokhin S.N. Natural radionuclides (^{238}U, ^{232}Th, ^{40}K) in the water of the northwestern Altai. World of Science, Culture, Education, 2007; **4**: 16–19 (in Russian)

[13] Tilton G.R. et al. Isotopic composition and distribution of lead, uranium, and thorium in a precambrian granite. Bulletin of the Geological Society of America,1956; **66** (9): 1131–1148.

[14] Rosholt J.N., et al. Evolution of the isotopic composition of uranium and thorium in Soil profiles. Bulletin of the Geological Society of America 1966; **77**(9): 987–1004.

[15] Bolsunovsky A.Y., Bondareva L.G. New data on the tritium content in one of the tributaries of the Yenisei River. Doklady Chemistry, 2002: 385(5): 714–717.

[16] Bolsunovsky A.Y., Bondareva L.G. Tritium in surface waters of the Yenisei River basin. Journal of Environmental Radioactivity, 2003; **66**: 285–294.

[17] Bondareva L.G., Zharovtsova S.A. Determination of tritium content in the environment. Vestnik KrasGU, ser. Analytical Chemistry, 2003; **2**: 127–128.

[18] Bondareva L.G. New data on the ecological state of the River Yenisei. Russian Journal of General Chemistry, 2010; **3**: 153–161.

[19] Bondareva L.G. Mechanisms of tritium transfer in freshwater ecosystems. Vestnik of National Nuclear Center. Republic Kazachstan, Bulletin of the National Nuclear Center. 2011; **1**: 10–23 (in Russian)

[20] Bondareva L. Natural occurrence of tritium in the ecosystem of the Yenisei river. Fusion Science and Technology, 2015; **60**(4): 1304–1307.

[21] Pirkko H. Radionuclide migration in crystalline rock fractures. Academic Dissertation, Helsinki, 2002.

[22] Salbu B., Krekling T. Characterisation of radioactive particles in the environment. Analyst, 1998; **123**: 843–850.

[23] Solatie D. Development and comparison of analytical methods for the determination of uranium and plutonium in spent fuel and environmental samples: Academic Dissertation. Helsinki, 2002. 63 p.

[24] Solatie I. D., Carbol P., Betti M. et al. Ion chromatography inductively coupled plasma mass spectrometry (IC-ICP-MS) and radiometric techniques for the determination of actinides in aqueous leachate solutions from uranium oxide. Fresenius Journal of Analytical Chemistry, 2000; **368**: 88–94.

[25] Colley S., Thomson J. Particulate/solution analysis of ^{226}Ra, ^{230}Th and ^{210}Pb in sea water sampled by in-situ large volume filtration and sorption by manganese oxyhydroxide. Science of the Total Environment, 1994; **155**: 273–283.

[26] Meece D.E., Benninger L.K. The coprecipitation of Pu and other radionuclides with $CaCO_3$. Geochimica et Cosmochimica Acta, 1993; **57**: 1447–1458.

[27] Cochran J. K., Livingston H. D., Hirschberg D. J. et al. Natural and anthropogenic radionuclide distributions in the northwest Atlantic Ocean. Earth and Planetary Science Letters, 1987; **84**: 135–152.

[28] Romantchuk A.U. Laws of sorption behavior of actinide ions in the mineral colloidal particles. Russian Journal of General Chemistry, 2010; **3**: 120–128.

[29] Petrova A.B. Sorption of Np and Pu in colloids particles of Fe(III) and Mn(IV) oxides in the present of gumic acids. Dissertation, 2007. 117 p (in Russian)

[30] Schnoor, J.L., Mossmann, D.J., Borzilov, V.A., Novitsky, M.A. et al. Mathematical model for chemical spills and distributed source runoff to large rivers. In: Schnoor, J.L. (Ed.), Fate of Pesticides and Chemicals in the Environment. New York: Wiley Interscience, Environmental Science and Technology Series, 1992. pp. 347–370.

[31] Belolipetsky B.M., Genova C.H. Calculating algorithm for definition of dynamics of suspended and bed sediments in channel. Computational Technologies, 2004; **2**: 9–25 (in Russian)

6

Modeling Agricultural Land Management to Improve Understanding of Nitrogen Leaching in an Irrigated Mediterranean Area in Southern Turkey

Ebru Karnez, Hande Sagir, Matjaž Gavan,

Muhammed Said Golpinar, Mahmut Cetin,

Mehmet Ali Akgul, Hayriye Ibrikci and Marina Pintar

Additional information is available at the end of the chapter

Abstract

Nitrogen (N) cycle dynamics and its transport in the ecosystem were always an attracting subject for the researchers. Calculation of N budget in agricultural systems with use of different empirical statistical methods is common practice in OECD and EU countries. However, these methodologies do not include climate and water cycle as part of the process. On the other hand, big scale studies are labor and work intensive. As a solution, various computer modeling approaches have been used to predict N budget and related N parameters. One of them is internationally established Soil and Water Assessment (SWAT) model, which was developed especially for modeling agricultural catchments. The aim of this study was to improve understanding of N leaching with simulation of agricultural land management (fertilization, irrigation, and plant species) in hydrological heavily modified watershed with irrigation-depended agriculture under Mediterranean climate. The study was conducted in Lower Seyhan River Plain Irrigation District (Akarsu) of 9495 ha in Cukurova region of southern Turkey. Intensive and extensive water and nitrogen monitoring data (2008–2014), soil properties, cropping pattern, and crop rotation were used for the SWAT model build, calibration, and validation of the model.

Keywords: crop management, irrigation, nitrogen balance, SWAT, modeling

1. Introduction

In arid and semiarid regions, freshwater resources are under the ever increasing pressure of many current issues such as population increase, economic development, climate change, and pollution [1]. Water quality is a major concern and expressed by its biological, chemical, physical, and aesthetic properties [2]. The water quality is determined by a number of factors such as electrical conductivity, pH, amount of salts, dissolved oxygen, levels of microorganisms, nutrients, heavy metals, quantities of pesticides, and herbicides [3]. These factors can lead to the problems (salinity, infiltration, toxicity, and nutrients), which are extensively present in many watersheds with irrigated agriculture [4–7].

Nitrogen leaching from agricultural land is a main pollutant in many countries in the world [7, 8]. In agricultural areas of the European Union (EU), fertilizer contribution as nonpoint source pollution to the surface water is estimated to be 55% [9]. The European Union Water Framework Directive (WFD) has issued important regulations in order to reduce the environmental impact of nitrogen due to agriculture and to keep water bodies in good quality state; based on the EU Drinking Water Directive (80/778/EEC), the accepted maximum admissible concentration for the nitrate was set as 50 mg l^{-1} [10].

On the other hand, nitrogen is an essential nutrient for adequate plant growth, and mostly used as type of fertilizer [11]. During the N cycle, it undergoes many processes in soil, water, and atmosphere level [12–14]. Nitrogen cannot be used directly by the plants and animals until it is converted into its available compounds and forms. Nitrate ions in soil are usually in dissolved form in the soil solution, and it can easily be lost to leaching as water moves through the soil profile due to the rapid dynamism [15, 16].

Understanding of nitrogen dynamics in the nature, nitrogen balance or nitrogen budget becomes more of an issue about prevention of environmental pollution and economic losses on a country basis. Nitrogen balance studies have been continued for over 170 years [17]. There are different ways of defining nitrogen budgets in empirical statistical methods, depending on the measurements and modeling. Calculation of N budget in agricultural systems by this way is a common practice in OECD and EU countries. This method does not include explaining the processes of nutrient cycle in the soil-plant-atmosphere system but follows statistical methodology at national and regional levels to determine nitrogen budget [18–20].

Measured nitrogen budgets in soil-plant-atmosphere level are based on the conservation of mass of nitrogen in the system. A previous study carried out [21, 22] aimed at evaluating nitrogen fluxes by measuring agronomic system in Akarsu Study Area in southern Turkey. As part of the findings, it was found that considerable amounts of nitrate are lost to drainage and shallow groundwater. During the study years, nitrogen budget calculations resulted in unaccounted values ranging from 40 to 60 kg N ha^{-1} [23].

As known, Mediterranean climate is characterized by mild rainy winters and hot dry summers [24]. Annual and interannual changes in dry and wet periods result in change of water balance and water level fluctuations especially in the areas where Mediterranean climate is dominating [25]. Based on the recent years' ongoing drought events and therefore water scarcity, irrigation scheduling and types need to be reevaluated. Recently, best management techniques such as drip irrigation [26] and rain water harvesting techniques [27] have been tried to put into practice in order to save both irrigation water and fertilizers. In the Mediterranean climate, irrigation is inevitable for maximizing the crop yield [28]. To increase crop yield and quality and at the same time to decrease the leaching below the rooting zone, managing nutrient concentrations in irrigation water is necessary, according to crop requirements [29].

Many tools are available to observe impacts of reduced irrigation and fertilization under agriculture best management practices (BMPs) scenario. Among those tools are different hydrological models capable of defining the nitrogen dynamics at the watershed level like AGNPS, AnnAGNPS, ANSWERS, ANSWERS-Continuous, CASC2D, DWSM, HSPF, KINEROS, MIKE SHE, APEX, and SWAT. And these are only a few of watershed modes, which are currently and commonly under the service of scientists and practitioners [30]. Soil and water assessment tool (SWAT) model is one of the tools developed to predict water and nutrient dynamics [31–34].

The aim of this study was to improve understanding of (a) the effects of bypass flows due to irrigation on the calibration of SWAT model, (b) irrigation return flow (IRF) and/or drainage generating processes, and (c) N leaching dynamics with simulation of agricultural land management (fertilization, irrigation, and plant species) under Mediterranean climate conditions.

2. Materials and methodology

2.1. Study area

The Akarsu Irrigation District (AID) study area is located in the Mediterranean coastal region, between 36°51′45″ and 36°57′35″ N latitudes, and 35°24′10″ and 35°36′20″ E longitudes in Turkey. The district covers an area of 9495 ha (irrigation area), and hydrological area is 11,308 ha in the Lower Seyhan Plain (LSP) and has been irrigated for over 60 years under conventional irrigation and drainage infrastructures. Until 1994, the national irrigation agency, i.e., State Hydraulic Works (DSI), was responsible for the management, operation, and maintenance of the district. Management of the irrigation and drainage system in the district was taken over by the water users in 1994. Akarsu Water User Association has been responsible for the irrigation management, operation, and repairing issues in the district since 1994. Irrigation water has been provided from Seyhan Dam (L6, L3, and L7 in **Figure 1**), in case of water shortage in the system during the peak irrigation season or if irrigation water is not diverted to the main irrigation canal through L6, then pumping station is activated and some water is diverted from Ceyhan River (Abdioglu Pumping Station, L9 in **Figure 1**). The irrigation water in Seyhan Dam has excellent water quality ($0.33 \leq EC \leq 0.50$ dS m^{-1}, EC = 0.43 dS m^{-1}). However, electrical conductivity (EC) of Ceyhan River is slightly higher than Seyhan ($0.41 \leq EC \leq 0.80$ dS m^{-1}, EC = 0.58 dS m^{-1}). The drainage water flows through open ditches along the downstream areas and finally discharges into the Mediterranean Sea.

In the study area, the Mediterranean climate is dominant, summers are hot and dry winters are mild and rainy. Precipitation is mostly in the form of rain (average of 659 mm) that usually falls during winter and spring [35]. Temperature in June, July, and August is very high (average 33.3°C); winter months are cool with reasonable temperatures (average 10.5°C) [36]. While the long-term (1929–2014) mean temperature is 27.4°C, the long-term mean total evaporation is about 1559 mm annually (coefficient of variation <27%). According to the long-term data, soil moisture and soil temperature regimes are defined as xeric and thermic by Ref. [37].

In the area, 1st April–30th September is defined as irrigation season (IS), while 1st October–1st April is defined as nonirrigation season (NIS). However, these dates may change a little by precipitation and climatic conditions.

The soils of Akarsu consist of 11 different soil series (Incirlik, Arikli, Yenice, Innapli, Arpaci, Canakci, Mursel, Ismailiye, Golyaka, Gemisure, and Misis). The model-related physical and chemical characteristics of these soil series are recorded from Ref. [37] and verified to be used in the SWAT model. As an example, only the data of six common soil series are given in **Table 1**. Arikli (29.5%), Incirlik (25.3%), and Yenice (12.2%) series cover 67% of the entire study area. Innapli (1.03%) and Mursel (0.7%) have got the minimum distributions.

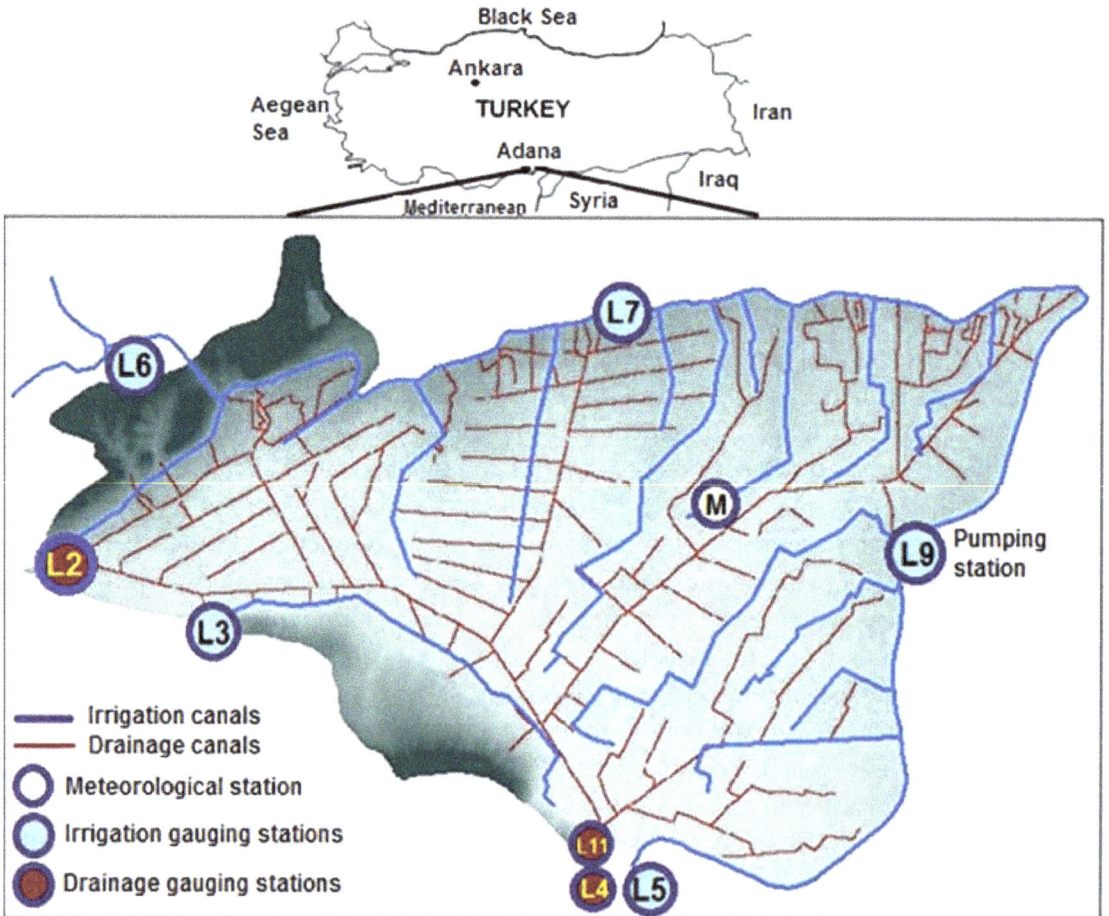

Figure 1. The Akarsu study area.

Soil series	Depth (cm)	Texture class[1]	Sand	Silt	Clay	Rock	BD[2]	OM[3]	AWC[4]	K_{sat}[5]
Incirlik	0–13	C	12	26	62	2.5	1.4	1.25	0.226	2.8
	13–78	C	16	24	60	2.5	1.6	0.62	0.233	0.55
	78–150	C	14	26	60	1.0	1.6	0.33	0.301	2.9
Arikli	0–13	SiC	8	29	63	3.5	1.3	1.25	0.268	0.63
	13–30	SiC	8	30	62	2.5	1.4	0.62	0.268	0.16
	30–57	SiC	4	29	67	1.5	1.4	0.33	0.287	0.4
	57–100		7	31	62			1.24		
	110–114		6	31	63			0.91		
	114–150		3	42	55			0.84		
Yenice	0–14	C	14	32	54	2.5	1.57	1.61	0.218	191.5
	14–32	C	12	30	58	1.8	1.59	1.21	0.259	328
	32–92	CL	14	30	56	1.0	1.48	0.80	0.219	729
	92–118		18	31	51			0.54		
Misis	0–24	C	25	23	52	1.5	1.63	1.61	0.212	3.33
	24–45	C	25	21	54	1.5	1.53	1.21	0.232	1.7
	45–64	SCL	23	21	56	1.7	1.49	0.93	0.247	1.7
	64–86		22	19	59		1.51	0.80		
	86–120		22	18	60			0.67		
	120–140		51	23	26			0.13		
Canakci	0–10	SL	25	47	28	1.5	1.51	1.37	0.208	23.9
	10–39	CL	21	55	24	1.8	1.34	1.17	0.171	16.8
	39–60	CL	29	39	32	1.6	1.58	1.50	0.157	11.5
	60–73		35	43	22			0.39		
	73–94		28	49	23			0.46		
	94–112		13	52	35			0.63		
	112–150		22	48	30			0.36		
Gemisure	0–21	C	2	26	72	1.5	1.45	1.53	0.15	2.4
	21–36	C	3	24	73	1.5	1.35	1.47	0.15	2.4
	36–78	C	3	22	75	1.8	1.39	1.34	0.15	2.4
	78–120		4	19	77			1.07		

[1] L, loam; C, clay; S, sand; Si, silt.
[2] Bulk density (g cm^{-3}).
[3] Organic matter (%).
[4] Plant available water capacity (mm H_2O mm soil $depth^{-1}$).
[5] Saturated hydraulic conductivity (mm h^{-1}).

Table 1. Soil properties for the Akarsu study area.

2.2. Database

The SWAT model input data, which is used in the project, is listed in **Table 2**. The 25 m resolution digital elevation model was derived by Akgul [38]. The chemical and physical properties of soils were gathered from Ref. [37], and these data were checked and verified with various measurements and laboratory analysis. Soil albedos and values of USLE were calculated by using the equations given in Ref. [39]. Soil series characteristics were interpreted and soil hydrologic group codes were assigned to each soil series based on the run-off generating characteristics. Daily irrigation return flow rates were determined by the data observed at the Inlet (L2, L11) and Outlet (L4) drainage monitoring stations. Nitrate concentrations were determined in water samples collected via automatic sampler located in L4 gauging site.

Data type	Resolution	Source	Description/properties
Topography (DEM)	25 m × 25 m	[38]	Elevation, slope, channel slopes, overland
Land cover/land use	10 m × 10 m	[35]	Land cover, land use classification
Soils	10 m × 10 m	[37]	Spatial soil variability, soil types, soil physical properties; bulk density, texture, saturated hydraulic conductivity classes, etc.
Drainage network		[35]	Drain spacing, length of cannels, drainage divides, etc.
Climate data		Adana State meteorological station and meteorological monitoring gage (L8)	Daily precipitation, temperature (max., min.), solar radiation, wind speed, relative humidity
Agricultural management practices		Farmer questionnaires in Akarsu and field surveys (face to face)	Planting, fertilizer application rates and timing, tillage, harvesting dates, irrigation water management and amount, etc.
Daily irrigation return flow rate (outlet)		1 monitoring and sampling station (L4 in **Figure 1**)	Daily flow (m^3 day^{-1})
Daily irrigation return flow rate (inlet)		2 monitoring and sampling stations (L2, L11)	Daily flow (m^3 day^{-1})
Daily irrigation return flow nitrate load (outlet)		1 monitoring and sampling station	Daily NO_3-N load (kg day^{-1})
Daily irrigation return flow nitrate load (inlet)		Two monitoring and sampling stations (L2, L11)	Daily NO_3-N load (kg day^{-1})

Table 2. Model input data and the sources.

2.3. Agricultural land management

The SWAT model has eight main components: hydrology, weather, sedimentation, soil temperature, crop growth, nutrients, pesticides, and agricultural management [30]. Watershed hydrology is affected by vegetation types, soil properties, geology, terrain, climate, land use practices, and spatial patterns of interactions among these factors [40].

The area is suitable for various agricultural productions with its favorable climatic and productive land conditions. Cropping pattern data have been assessed since 2006, and the likely crop rotation has been decided for the modeling practices. According to the data, land use and cropping pattern varied from year to year depending on the market and cultivation conditions. Based on the assessments, we have set five different crop rotations plus fruit orchards and citrus plantations (**Table 3**), which have been well adopted by the farmers in the region. Based on the recent years' evaluation, the main crops in the area were wheat, corn, citrus, cotton, and vegetables (**Table 3**). Agricultural management practices were determined based on the current surveys carried out at the local field and farmers' level.

Year	Soil tillage and crop growing period	Crops	Inorganic nitrogen fertilizer (kg elemental N ha^{-1})	Irrigation water (mm)
Rotation 1				
1	16th Mar.–16th Sep.	C1[1]	385	1168
1/2	20th Nov.–07th June	WW[2]	230	383
2/3	15th June–10th Oct.	S2[3]	120	870
3	16th Mar.–16th Sep.	C1[1]	385	1168
3/4	20th Nov.–1st June	WW[2]	230	383
4	15th June–10th Oct.	S2[3]	120	870
Rotation 2				
1	15th June–10th Oct.	S2[3]	120	870
2	16th Mar.–16th Sep.	C1[1]	385	1168
2/3	20th Nov–07th June	WW[2]	230	383
3	15th June–10th Oct.	S2[3]	120	870
4	16th Mar.–16th Sep.	C1[1]	385	1168
Rotation 3				
1	15th Mar.–15th Oct.	Co[4]	290	1535
2	15th Apr.–10th Sep.	P1[5]	210	1068.33
3	15th Mar.–15th Oct.	Co[4]	290	1535
4	16th Mar.–16th Sep.	C1[1]	385	1168.33
Rotation 4				
1	15th June–25th Oct.	P2[6]	210	800
2	16th Mar.–16th Sep.	C1[1]	385	1168.33
2/3	20th Nov.–07th June	WW[2]	230	383.33
3	15th June–25th Oct.	P2[6]	210	800
4	15th Mar.–15th Oct.	Co[4]	290	1535
4/1	20th Nov.– 07th June	WW[2]	230	383.33

Year	Soil tillage and crop growing period	Crops	Inorganic nitrogen fertilizer (kg elemental N ha^{-1})	Irrigation water (mm)
Rotation 5				
1	20th June–30th Oct.	C2[7]	330	858.33
2	16th Mar.–16th Sep.	C1[1]	385	1168.33
2/3	20th Nov.–07th June	WW[2]	230	383.33
3	20th June–30th Oct	C2[7]	330	858.33
4	15th Mar.–15th Oct.	Co[4]	290	1535
4/1	20th Nov.–07th June	WW[2]	230	383.33
Orchards and citrus[+]				
Perennial	15th Mar.–8th Oct.	Orchards	250	1238.33
Perennial	1st Oct.–27th Sep.	Citrus	335	1040

[1] C1, first crop corn.
[2] WW, winter wheat.
[3] S2, second crop soybean.
[4] Co, cotton.
[5] P1, first crop peanut.
[6] P2, second crop peanut.
[7] C2, second crop corn.
[+] All kinds of operations done to orchards and citrus between these dates.

Table 3. Agricultural land management crop rotations used in the model.

The proportion of this land use type in the hydrological model area (11,308 ha) is: AGRL (Agricultural Area) (64.56%), ORAN (Citrus) (21.49%), ORCD (Orchards) (1.74%), WPAS (Winter Pastures) (9.20%), URMD (Settlement area (Medium Density)) (1.64%), and URLD (Settlement area (Low Density) (1.36%)). The agricultural areas in the study area contain various annual crops such as first crop corn, second crop corn, winter wheat, first crop soybean, second crop soybean, peanuts, and cotton.

2.4. SWAT model description

The soil and water assessment tool is one of the recent models, known as a catchment area or watershed scale model, developed by Arnold et al. [31] and improved in the last 30 years [41]. It is a semidistributed hydrological model, which is a physically based, long period of simulation, lumped parameter, and derived from agriculture management systems models such as CREAMS, EPIC, and GLEAMS [41, 42]. The model separates selected basin to subbasins and hydrologic response units (HRU) comprised of identical hydrological properties such as land use, soil, and slope [43]. SWAT is an efficient tool to predict the impact of nitrogen cycle and land management practices on water, sediment, nutrient, and pesticide with the ArcSWAT module [44]. The nitrogen cycle can be represented by the SWAT model in the soil profile and shallow aquifer. SWAT comprises two pools that are inorganic forms of nitrogen (NH_4^+ and NO_3^-) and three pools that are organic forms of nitrogen in the soil [45–47]. Nitrate and organic N into the nitrogen cycle, N removal from soil to water sources, and amounts of NO_3-N included in lateral flow, runoff, and percolation can also be represented by the SWAT model [45]. The SWAT model could sufficiently predict sediment and nutrient statuses as well as tile drainage NO_3-N losses [48, 49].

The prediction of land management practices is important as well as nitrogen cycle to provide the progress of future socioeconomic stability and sustainable use of natural resources and to search the impact of human activities on a given basin [50, 39]. SWAT has a capability to estimate the effects of land management practices on sediment, water, and agricultural chemical yields in large complex watersheds with varying soils, land use, and management conditions over a long-term time [43, 51–54].

2.5. Calibration process of the base model

Calibration and validation are key processes in reducing the uncertainty and increasing user experience in its predictive results, making the software a practical model and leading to user competence.

The adjustment of model parameters is described as calibration. These parameters are associated with checking results toward observations to assure the same response in time [55]. A number of calibration techniques, comprising manual calibration method and automated method, improved for the SWAT model [39]. The model calibration is done manually and finalized by SWAT-CUP (Calibration and Uncertainty Programs). SWAT-CUP is an interface known as "automated model calibration method" that was improved for SWAT to connect with a link between the input and output of a calibration program and the model [56]. The SUFI-2 algorithm was used for sensitivity analysis, model calibration, and validation process. The warm up period was set for 1 year.

The calibration of SWAT is completed in three phases [39]. The first phase is the determination of most sensitive parameters (such as Alpha_Bf, Canmx, Ch_K2, Ch_N, Cn2, Esco, Gw_Delay, Gw_Revap, Gwqmn Surlag for flow, and Nperco, All CMN, Hlife_Ngw for nitrate-nitrogen) [57]. The second phase is model calibration with use of statistical methods such as Pearson coefficient of correlation (R^2), Nash-Sutcliffe efficiency (ENS), and percent bias (PBIAS). The final phase is validation process for hydrological calibration and nitrogen calibration of the model.

Validation, known as the part of simulation, can be done without modifying any parameter values adjusted during calibration for a different time series to input data and also for the same time period at a different spatial location [58]. In this study, daily measured values of irrigation and irrigation return flows, and also nitrate loads for the year of 2008 were used for the warm up period. SWAT was calibrated with daily values over a 4-year period from 2009 to 2012 for hydrological years and used daily values for nitrogen. The 2-year time period from 2013 to 2014 was used for validation of hydrology and nitrogen.

3. Results and discussion

3.1. Calibration of drainage flows

Calibration process of the model used in this specific research was first completed with hydrologic calibration and followed by the drainage nitrogen. In general, calibration and validation of water quality models are typically performed with data collected at the outlet of a watershed to be able to assess possible pollution risks. In Akarsu, daily measured data were used during the model processes. The most sensitive parameters for hydrologic calibration process were SURLAG, GW_Delay, Revapmn, GW_Revap, and Esco (**Table 4**), while Nperco, Cmn, Hlife, and Ngw are the sensitive ones for nitrogen calibration.

Parameter	File extensions	Explanation
Alpha_BF	.gw	Base flow recession factor, days
GW_DELAY	.gw	Groundwater delay, days
SURLAG	.bsn	Surface runoff lag coefficient, days
ESCO	.bsn	Soil evaporation compensation factor
GWQMN	.gw	Threshold depth for ground water flow to occur, mm
GW_REVAP	.gw	Groundwater "revap" coefficient
Ch_N2	.rte	Manning's n
SOL_AWC	.sol	Available water capacity
CN2	.mgt	SCS curve number, antecedent moisture condition II, for crop land use
REVAPMN	.gw	Threshold water depth in shallow aquifer for percolation to deep aquifer to occur
CH_K2	.rte	Effective hydraulic conductivity in main channel alluvium
NPERCO	.bsn	Nitrate percolation coefficient
HLIFE_NGW	.gw	Half-life of nitrate in shallow aquifer (days)
CMN	.bsn	Rate factor for humus mineralization of active organic nutrients (N).
AI1	.wwq	Fraction of algal biomass that is nitrogen (mg N mg alg^{-1}).

Table 4. SWAT input parameters for river flow and nitrogen calibrations.

Based on the model outputs, the SWAT model is reliable enough to be used in nonnatural catchments such as Akarsu Irrigation District where drainage network is not topography-driven but man-made. Additionally, hydrologic water dynamics such as inflows, outflows, and the whole water balance are well defined since 2006. The area is affected by routine agricultural management activities, i.e., irrigation and fertilization in specific.

Three recommended quantitative statistics, determination (R^2), Nash-Sutcliffe efficiency (NSE), and PBIAS, in addition to the graphical techniques for visual examination have been used to assess the hydrologic model performance [59], i.e., model calibration and validation. These performance indicators of the model (R^2, NSE, and PBIAS) during calibration period of 2009–2012 have been found as 0.62, 0.57, and 6.3, respectively (**Table 5**). Typically, values of R^2

greater than 0.50, while values of NSE between 0.0 and 1.0, and values of PBIAS ±25% for streamflow calibration are generally considered as acceptable levels [59]. In addition, model validation was made by utilizing the daily data for 2013 and 2014 period. The performance statistics for the validation period were 0.67, 0.59, and −10.04 for R^2, NSE, and PBIAS, respectively (**Table 5**).

Variable	R^2	NSE	PBIAS
Calibration (2009–2012)			
Daily drainage flow	0.62	0.57	6.3
Daily nitrogen loss	0.47	−0.63	88.1
Validation (2013–2014)			
Daily drainage flow	0.67	0.59	−10.04
Daily nitrogen loss	0.50	−0.20	72.9

Table 5. Objective function statistics for drainage flow and nitrogen in drainage.

Descriptive statistics for observed and simulated (calibration and validation) were resented in **Table 6**, indicating that model performance was satisfactory with the mean values of 3.51, 2.98 $m^3 s^{-1}$ for calibration period and 2.71 and 2.98 for validation period. Similarly, other descriptive statistics for observed and simulated flow values were in good agreement.

The visual examination of observed versus predicted drainage flows for the calibration (**Figure 2**) and validation periods (**Figure 3**) indicated adequate calibration and validation. Therefore, SWAT simulations and observed data were in good agreement visually and statistically. SWAT-CUP automatic calibration results for the sensitive parameters were presented in **Table 7**. These parameters are reasonable enough to accept performance of the model [56] in a well-defined agricultural catchment of Akarsu where anthropogenic factors affecting hydrological processes are very preponderant.

Because the study area is under irrigation in dry periods of the year, it was necessary to consider irrigation amounts of field and horticultural crops grown in the region. Therefore, during the calibration period, irrigation requirements of the crops were estimated by using universal reference evapotranspiration method of Penman-Monteith. Then, using the crop coefficients of FAO [60], net irrigation requirements of irrigated crops were obtained and used in management files as a model input. For the calibration, the created base model with net irrigation amounts and routine fertilizer rates were saved in crop rotations. The actual irrigation bypass flows were determined through running different simulations by adapting calibrated SWAT parameters given in **Table 7**. Finally, it was determined that 40% of the total diverted irrigation water to the district at any time was directly draining into the drainage system as bypass flow.

3.2. Calibration of nitrogen in drainage water

After calibrating the hydrologic part of the model with a successful performance, nitrate simulation was confidently applied with the appropriate water parameters. All the nitrogen inputs were incorporated in the management files as fertilizer, water, and soil point sources.

Average daily NO_3-N loads (kg day^{-1}) were selected as water quality parameter and calculated based on daily discharge data (m^3 day^{-1}) at L4 gauging station (outlet).

	Calibration period (2009–2012)		Validation period (2013–2014)	
	Observed	Simulated	Observed	Simulated
		Drainage flows (m^3 s^{-1})		
Mean	3.51	2.98	2.71	2.98
Median	3.48	2.69	2.74	2.62
Mode	1.04	1.86	1.23	2.35
Standard dev.	2.02	1.93	1.45	1.46
Kurtosis	5.54	10.43	−1.01	−0.62
Skewness	1.70	1.96	0.33	0.38
Minimum	0.73	0.09	0.58	0.31
Maximum	14.06	18.65	6.05	7.27
CV%	58	65	54	49
		Nitrogen in drainage (kg d^{-1})		
Mean	1810.2	443.1	938.7	552.6
Median	1376.4	243.9	774.0	457.6
Mode	1775.5	229.3	–	1135.0
Standard dev.	1659.0	639.9	608.9	533.2
Kurtosis	15.8	21.9	5.8	57.8
Skewness	3.4	4.0	2.0	5.8
Minimum	143.4	10.4	64.9	39.7
Maximum	14024.7	6403.0	4826.0	7599.0
CV%	92	144	65	96
Sample size (n)	1461		730	

Table 6. Descriptive statistics of drainage flows and nitrogen loads in drainage for observed and simulated during calibration and validation periods.

Objective function statistics, R^2, NSE, and PBIAS in specific, for nitrogen in drainage were defined as 0.47, −0.63, and 88.1% for the calibration and 0.50, −0.20, and 72.9% for validation, respectively (**Table 5**). As indicated by Moriasi et al. [59], the PBIAS ±70% for N is accepted as a performance criteria.

Average daily NO_3-N loads (kg day^{-1}) of the selected water quality parameter was also calculated based on daily discharge data (m^3 day^{-1}) at L4 gauging station at the outlet of the district. Based on the graphical presentation in **Figures 4** and **5**, overlapping of the both measured and calibrated lines for N cannot be considered as perfect because nature and dynamics of N in the whole system, even though the statistics are reasonably acceptable. Similar underestimation with the data of only 2009 and 2010 was also recorded in the same location [38]. It is important to point out that calibration and validation of the model are sensitive to time periods, instead of using daily data, monthly data were more suitable to modeling purpose of N [61].

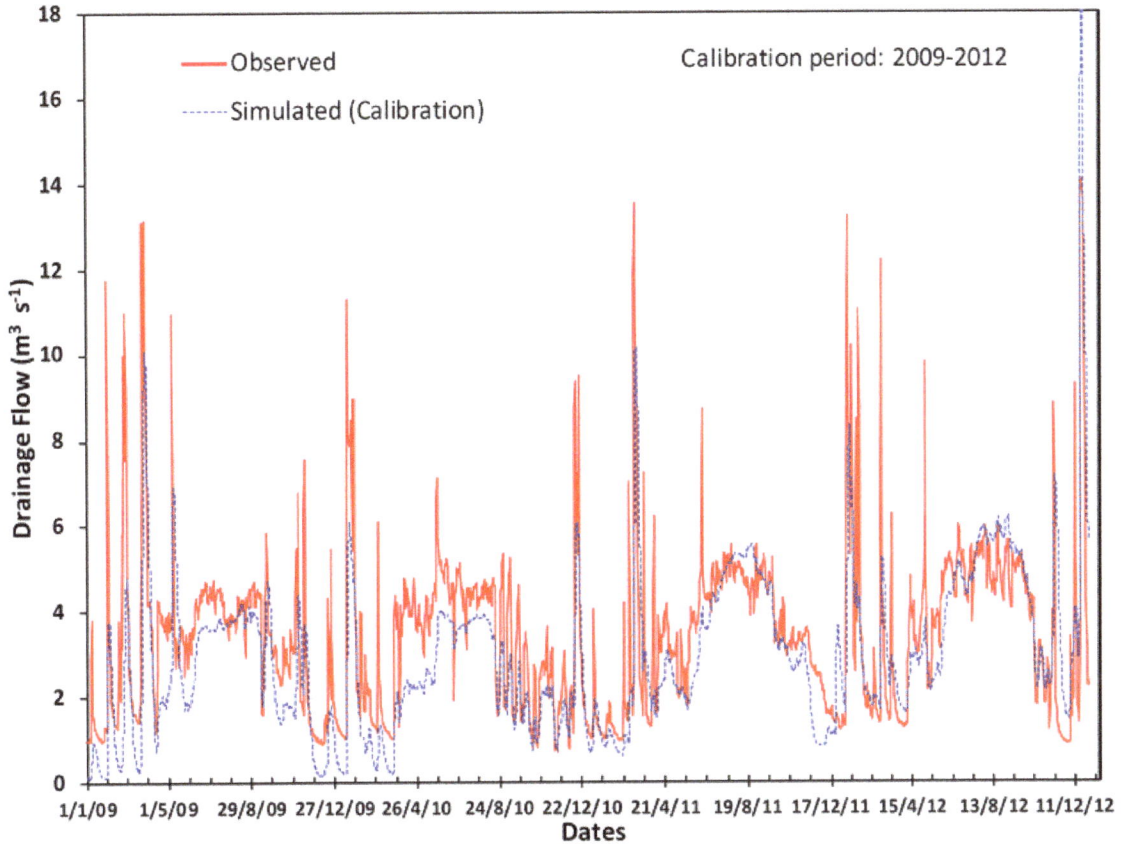

Figure 2. Daily drainage discharge ($m^3 s^{-1}$) calibration for the Akarsu catchment outlet L4.

This basin is not natural instead it is a man-made hydrologically well-defined area in a semiarid Mediterranean region where it is subjected to intensive irrigation and fertilizer applications by anthropogenic activities. Imported N loads by irrigation water, rainfall, and inorganic fertilizer inputs make the calibration and validation difficult and relatively weak. There are three district-specific conditions in the area to be pointed out for nitrogen and nitrogen balance: the canals being open despite high rates of ET, irrigation also taking place outside of irrigation season, and the possible loses of irrigation water to drainage.

In terms of management practices, there are two planting seasons in a year; the crop rotations used in the model include all the planting and harvesting dates. Except for perennial crops, the crop pattern (land use) varies from year to year. The model permits use of only one land use map in HRU delineation; for this reason, rotation calendars were made to be utilized within the model. Farmer behavior and knowledge are diverse, and the use of nitrogen fertilizers and irrigation is intense.

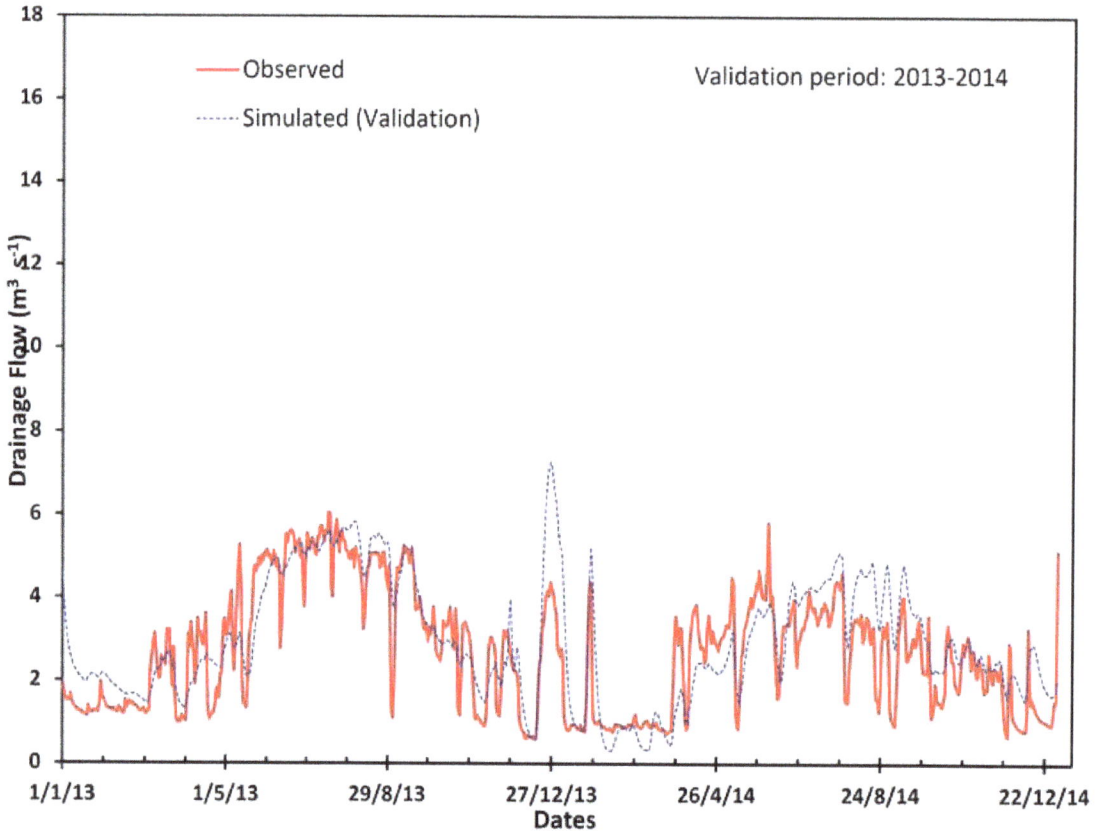

Figure 3. Daily drainage discharge ($m^3\ s^{-1}$) validation for the Akarsu catchment outlet at L4.

Parameter	Default	Range	Calibrated values
CN2	83	35–98	73.9
Alpha_BF	0.048	0–1	0.55
GW_Delay	31	0–500	36.08
Gwqmn	1000	0–5000	4187.5
Surlag	4	1–24	0.42
Esco	0.95	0–1	0.837
Revapmn	750	0–1000	488.75
Ch_K2	0	−0.01 to 500	378.75
Gw_Revap	0.02	0.02–0.2	0.089
Ch_n2	0.014	−0.01 to 0.3	0.266

Table 7. Sensitive hydrologic model parameters for SWAT.

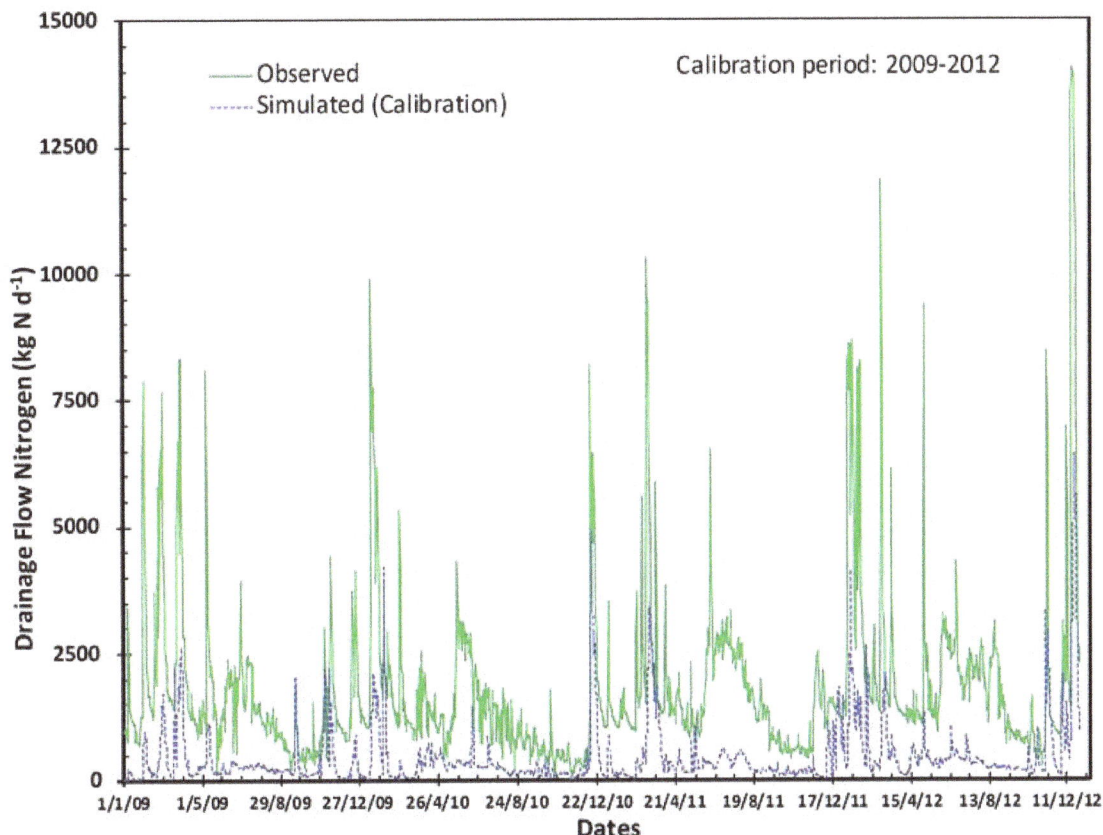

Figure 4. Nitrogen load (kg day^{-1}) at L4 (outlet) calibration and validation period on monthly level.

3.3. Nitrogen balance

Nitrogen calibration was carried out on daily basis. Average daily NO$_3$-N loads (kg day^{-1}) of selected water quality parameter were calculated based on daily discharge data (m^3 day^{-1}) at L4 gauging station. **Table 5** and **Figures 4** and **5** created for nitrogen did not show a strong relationship between measured and simulated values. One of the main reasons is that for hydrologic reasons inclusion of the two hilly pasture areas (**Figure 1**) into the 9495 ha hydrologically well-defined Akarsu irrigation district by extending the area to 11,308 ha. Therefore, when the actual N inputs were distributed in a larger area the prediction became lower. Also, since the soils are climatically suitable to nitrification, greater amount of nitrogen especially from the inorganic fertilizers may be quickly transformed to nitrate in a very short time period and leached to the drainage [62]. As also discussed by Abbaspour et al. [56], amount of nitrogen fertilizer leached below the root zone, which is 0–90 cm in the study, is under-estimated. In addition, fertilizer application level may be higher than that of the recorded from our three consecutive survey data. Therefore, it may cause higher measured NO$_3$ concentrations in drainage. Overall, since the irrigated area is under very intensive agricultural management practices including irrigation and very dynamic fertilization, it is quite possible to underestimate the N leaching to the drainage. For example, SWAT model prediction was very successful for calibration (and validation) of rivers accounting the dynamics of nitrate transport [56].

Nitrogen balance variables are given in **Table 8**. The sums of nitrate nitrogen leached from the soil profile in kg NO_3–$N(NO3L)$ and N uptake by plants (NUP) from 2009 to 2014 are reasonably in agreement with the amount of applied nitrogen (N APP). The remaining inputs in the so-called man-made research area are coming from the N content of irrigation water, rainfall, mineralization of soil organic matter, and transforms of N forms into readily available NH_4^- and NO_3. Based on the climatic conditions, amount of rainfall, thus leaching to drainage, and groundwater, varies year to year. For example, in 2013, total rainfall was 349 mm, which was the lowest figure among the other years of the study (ranged 349–951 mm). The reflection of this unusual rainfall was clearly performed in **Figure 5**, which is for the simulation period. **Figure 4** clearly indicates that impacts of rainfall in winter and irrigation applications in

Year	Nitrogen balance variables (kg N ha^{-1} year^{-1})		
	N_APP[1*]	NO3L	NUP
2009	329.2	196.8	270.0
2010	368.1	212.8	228.3
2011	310.9	234.7	181.3
2012	368.1	256.3	175.1
2013	329.2	159.2	277.7
2014	368.1	249.3	254.6

[*] N_APP, NO3L, and NUP stand for applied, leached, and taken-up nitrogen at the catchment level.

Table 8. Temporal variability of nitrogen balance by SWAT modeling for the Akarsu region (2009–2014).

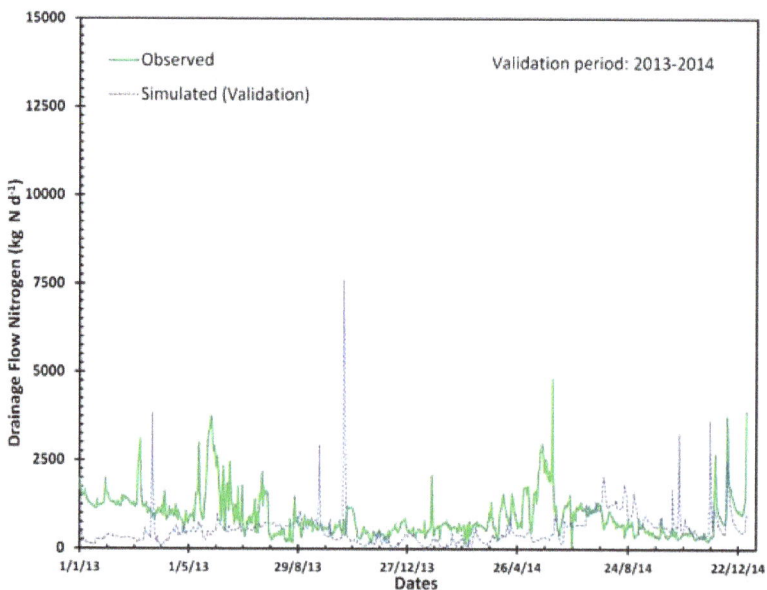

Figure 5. Nitrogen load (kg day^{-1}) at L4 (outlet) calibration and validation period on monthly level.

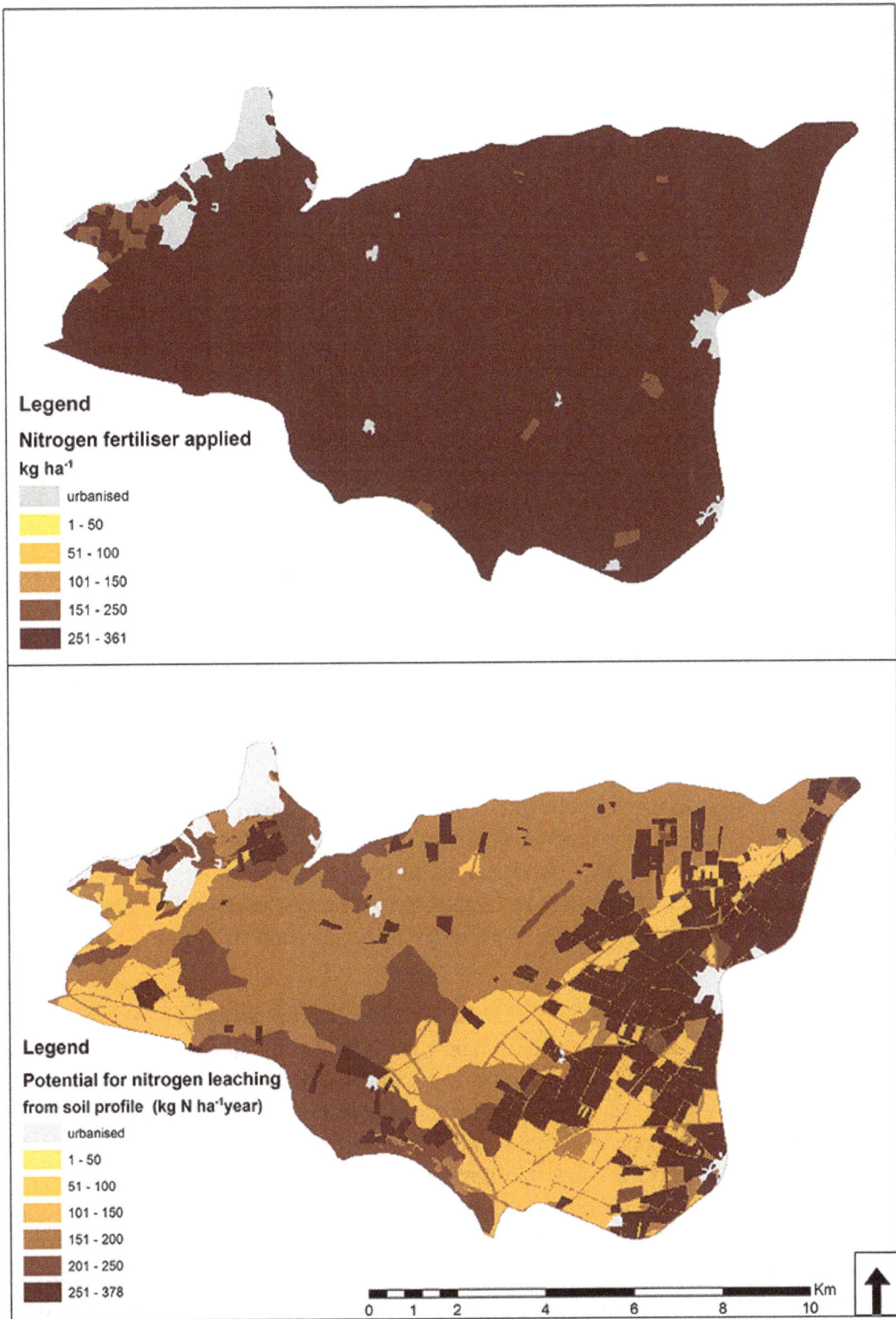

Figure 6. Comparison between average nitrogen fertilizers applied (kg ha^{-1}) and potential for nitrogen leaching (kg ha^{-1}) below the bottom of the soil profile in Akarsu study area in the period between 2009 and 2014.

summer are the most important drivers of the N leaching. Conflicting performance ratings of N calibration seen in **Figures** 4 and 5 might be attributed to above mentioned two drivers. In addition, routine fertilizer applications are exceedingly high than the recommended levels, i.e., 380 kg N ha^{-1} is applied to corn while only 240 kg N ha^{-1} is the expert recommendation for corn in the region [63]. This results in high potential for nitrogen leaching (**Figure 6**).

4. Conclusions

Distributed watershed models are known as the very powerful tools both for scenario development and for simulating the effects of watershed dynamics management on soil and water resources. This study was aimed to improve understanding of (a) the effects of bypass flows due to irrigation on the calibration of the SWAT model, (b) irrigation return flow and/or drainage generating processes, and (c) N leaching dynamics with simulation of agricultural land management (fertilization, irrigation, and plant species) under the Mediterranean climate conditions. To this aim, the research was conducted in an irrigated agricultural catchment of Akarsu irrigation district. Visual examination of data used in modeling has indicated that drainage flows and nitrogen-leaching processes are not governed by the natural processes in the catchment but mostly by anthropogenic activities.

Model calibration and validation were carried out to determine the most sensitive and appropriate parameter values for the drainage flows generated by the agricultural catchment. Although daily flow data were used in modeling, quantitative model performance evaluation statistics (R^2, NSE, and PBIAS) revealed clearly that the calibrated SWAT model produced rather satisfactory simulation results at the catchment outlet in wet, average, and dry years. In the irrigated catchment, irrigation water losses directly from irrigation channels to drainage ditches, i.e., bypass flows, has direct influence on calibrating hydrologic part of the SWAT model. In this case, the SWAT model findings helped us to highlight that almost 40% of diverted irrigation water has been recklessly squandered in the irrigation scheme. It is almost impossible to quantify bypass flow magnitudes in such irrigation system without using any modeling tools.

Furthermore, modeling exercises showed that the SWAT model run results were sensitive on crop rotations due to the fact that runoff by precipitation and irrigation applications are affected by the land use and land cover types. Contrary to the expectations, daily nitrate modeling results were not able to yield rather satisfactory model performance statistics, indicating that simulated daily nitrogen loads data in drainage were not sufficiently matched with the measured ones. Visual evaluation of measured and simulated nitrogen graphs showed implicit signals that measured nitrogen data might involve some inherent uncertainties and irregularities at the catchment level. Based on the findings, as highlighted in the literature [59], we concluded that model performance can be improved to some extent by increasing the time step from daily to monthly or yearly level for the nitrogen data with involves inherent uncertainties. These uncertainties should be considered when calibrating, validating, and evaluating watershed models because of differences in inherent uncertainty between measured flow, sediment, and nutrient data.

Improved fertilization practices are not only necessary for farmer's economy but also crucial for preserving soil and water resources. In recent years, especial soil analysis in the study area became a very useful tool for fertilizer subsidizes and expert recommendations. However, recommendations can not only be related to and designed by the soil analysis, it should be comprehensively evaluated in a broader environment. At this stage, a suitable model performance enables modeling more sensitive management practices like the fertilizer rates.

Acknowledgements

This work is a results of cooperation between researchers included in Slovenia-Turkey bilateral project financed by Slovenian Research Agency (ARRS) and the Scientific and Technological Research Council of Turkey (TUBITAK, Project No: 213O057).

Author details

Ebru Karnez[1], Hande Sagir[2], Matjaž Gavan[3]*, Muhammed Said Golpinar[4], Mahmut Cetin[4], Mehmet Ali Akgul[5], Hayriye Ibrikci[2] and Marina Pintar[3]

*Address all correspondence to: matjaz.glavan@bf.uni-lj.si

1 Faculty of Agriculture, Cukurova University, Adana, Turkey

2 Soil Science and Plant Nutrition Department, Cukurova University, Adana, Turkey

3 Department for Agronomy, Biotechnical Faculty, University of Ljubljana, Ljubljana, Slovenia

4 Department of Irrigation Engineering, Faculty of Agriculture, Cukurova University, Adana, Turkey

5 The Sixth Regional Directorate of State Hydraulic Works (DSI), Adana, Turkey

References

[1] Wheater HS, Mathias SA, Li X, editors. Groundwater modelling in arid and semi-arid areas. International Hydrology Series, Cambridge University Press; New York, USA; 2010.

[2] Office of Environment and Heritage, 2015. http://www.environment.nsw.gov.au/water/waterqual.htm

[3] National Marine Sanctuaries, 2011. floridakeys.noaa.gov/ocean/waterquality.html

[4]   FAO Corporate Document Repository, 1985. http://www.fao.org/docrep/003/t0234e/T0234E01.htm#ch1.3

[5]   Koc C. A study on the pollution and water quality modeling of the river Buyuk Menderes, Turkey. Clean-Soil, Air, Water. 2010; 38 (12): 1169–1176.

[6]   Volk M, Liersch S, Schmidt G. Towards the implementation of the European Water Framework Directive? Lessons learned from water quality simulations in an agricultural watershed. Land Use Policy. 2009; 26: 580–588.

[7]   Du B, Saleh A, Jaynes DB, Arnold JG. Evaluation of swat in simulating nitrate nitrogen and atrazine fates in a watershed with tiles and potholes. American Society of Agricultural and Biological Engineers. 2006; 49 (4): 949–959.

[8]   O'Shea L, Wade A. Controlling nitrate pollution: an integrated approach. Land Use Policy. 2009; 26:799–808.

[9]   Kersebaum KC, Steidl J, Bauer O, Piorr H-P. Modelling scenarios to assess the effects of different agricultural management and land use options to reduce diffuse nitrogen pollution into the river Elbe. Physics and Chemistry of the Earth, Parts A/B/C. 2003; 28: 537–545. DOI: 10.1016/S1474-7065(03)00090-1

[10] Croll BT, Hayes CR. Nitrate and water supplies in the United Kingdom. Environmental Pollution. 1988; 50: 163–187.

[11] Marschner H.Marschner's Mineral Nutrition of Higher Plants (Third Edition Book). Academic Press, Elsevier; Oxford, UK; 2002. ISBN: 978-0-12-384905-2.

[12] Zhu ZL, Chen DL. Nitrogen fertilizer use in China-Contributions to food production, impacts on the environment and best management strategies. Nutrient Cycling in Agroecosystems. 2002; 63 (2): 117–127.

[13] Raun WR, Schepers JS. Nitrogen management for improved use efficiency. In JS Schepers and WR Raun, editors. Nitrogen in agricultural systems. American Society of Agronomy Monograph no. 49. American Society of Agronomy, Madison, WI, USA; 2008. pp. 675–695.

[14] Smil V. Nitrogen cycle and world food production. World Agriculture. 2011; 2: 9–13.

[15] Kladivko EJ, Scoyoc GEV, Monke EJ, Oates KM, Pask W. Pesticide and nutrient movement into subsurface tile drains on a silt loam soil in Indiana. Journal of Environmental Quality. 1991; 20: 264–270.

[16] Ng HYF, Tan CS, Drury CF, Gaynor JD. Controlled drainage and subirrigation influences tile nitrate loss and corn yields in a sandy loam soil in Southwestern Ontario. Agriculture Ecosystems and Environment. 2002; 90: 81–88. DOI: 10.1016//S0167-8809(01)00172-4

[17] Meisinger JJ, Schepers JS, Raun WR. Crop nitrogen requirement and fertilization. In JS Schepers, et al. editors. Nitrogen in agricultural systems. Agronomy Monograph. 49. ASA, CSSA, and SSSA, Madison, WI; 2008. pp. 563–612.

[18] Kopinski J, Tujaka A, Igras J. Nitrogen and phosphorus budgets in Poland as a tool for sustainable nutrients management. Acta Agriculturae Slovenica. 2006; 87 (1): 173–181.

[19] Ozbek FS, Leip A. Estimating the gross nitrogen budget under soil nitrogen stock changes: a case study for Turkey. Agriculture Ecosystems & Environment. 2015; 205: 48–56.

[20] EC. Development of agri-environmental indicators for monitoring the integration of environmental concerns into the common agricultural policy. Communication from the Commission to the Council and the European Parliament. COM, 508 final. Commission of the European Communities, Brussels. 2006. Available from: http://eur-lex.europa.eu/LexUriServ/LexUriServ.do?uri=COM2006.0508:FIN:EN:PDF

[21] Cetin M. Nitrogen input-output mass balance study results in the selected Mediterranean countries. International Seminar: diagnosis and control of diffuse pollution in mediterranean irrigated agriculture, Zaragoza, Spain, 20–21 October 2010 (http://www.iamz.ciheam.org/qualiwater/ZaragozaSeminar/INDEX%20OF%20PRESENTATIONS)

[22] Karnez E, Evaluation of nitrogen budget inputs and outputs in wheat and corn growing areas in lower Seyhan plain (Ph.D. Dissertation). Adana, Turkey: Çukurova University (in Turkish); 2010.

[23] Ibrikci H, Cetin M, Karnez E, Topcu S, Kirda C, Ryan J, Oguz H, Oztekin E, Dingil M. Plant contribution to the nitrogen budget under irrigation in the Cukurova region of Southern Turkey. In: World Soils Congress Australia; 1–6 August 2010.

[24] Lionello P, Malanotte-Rizzoli P, Boscolo R, Alpert P, Artale V, Li L, Luterbacher J, May W, Trigo R, Tsimplis M, Ulbrich U, Xoplaki E. The Mediterranean climate: an overview of the main characteristics and issues. Developments in Earth and Environmental Sciences. 2006; 4: 1–26.

[25] Beklioğlu M, Çakıroğlu A, Tavşanoğlu N, Levi E, Özen A, Coppens J, Balaman SB, Jeppesen E. Impact of climate and nutrient enrichment on ecology of shallow lakes of Turkey using multiple approaches. In: Ecology and evolutionary biology symposium (EEBST 2015); Ankara, Turkey, 05–07 August 2015.

[26] FAO-ISRIC. 1990. Guidelines for profile description. 3rd Edition. Rome.

[27] Ramblinrobert. Irrigating with rainwater in a Mediterranean climate. Urban Agroecology. 2012. Available from: urbanagroecology.org/2012/12/05/irrigating-with-rainwater-in-a-mediterranean-climate

[28] Cavero J, Barros R, Sellam F, Topcu S, Isidoro D, Lounis A, Ibrikci H, Cetin M, Williams JR, Aragüés R. APEX simulation of best irrigation and n management strategies for off-site n pollution control in three Mediterranean irrigated watersheds. Agricultural Water Management. 2012; 103: 88–99.

[29] Hagin J, Lowengart A. Fertigation for minimizing environmental pollution by fertilizers. Fertilizer research. 1996; 43 (1): 5–7.

[30] Borah DK. Water resources models. In: K.C. Ting, D.H. Fleisher, and L.F. Rodriguez, editors. System analysis and modeling in food and agriculture, Encyclopedia of Life Support Systems; EOLSS Publisher, UNESCO; Oxford, UK; 2009.

[31] Arnold JG, Srinivasan R, Muttiah RS, Williams JR. Large area hydrologic modeling and assessment: Part I. Model development. Journal of American Water Resources Association. 1998; 34(1): 73–89.

[32] Panagopoulos Y, Makropoulos C, Mimikou M. A multi-objective decision support tool for rural basin management. In: http://www.iemss.org/sites/iemss2012/proceedings.html. International Environmental Modelling and Software Society (iEMSs). International Congress on Environmental Modelling and Software. Managing Resources of a Limited Planet, Pathways and Visions under Uncertainty, Sixth Biennial Meeting; Leipzig, Germany; 2012.

[33] Ertürk A, Ekdal A, Gürel M, Karakaya N, Guzel C, Gönenç E. Evaluating the impact of climate change on groundwater resources in a small Mediterranean watershed. Science of the Total Environment. 2014; 499: 437–447. DOI: 10.1016/j.scitotenv.2014.07.001

[34] Savé R, Herralde FD, Aranda X, Pla E, Pascual D, Funes I, Biel C. Potential changes in irrigation requirements and phenology of maize, apple trees and alfalfa under global change conditions in Fluvià watershed during XXIst century: results from a modeling approximation to watershed-level water balance. Agricultural Water Management. 2012; 114: 78–87.

[35] Çetin M, İbrikçi H, Berberoğlu, S, Gültekin U, Karnez E, Selek B. Analysis and optimization of irrigation efficiencies in order to reduce salinization impacts in intensively used agricultural landscapes of the semiarid Mediterranean Turkey (MedSalin). Project Final Report, Funded by TÜBİTAK-BMBF, Program Code: 2527, Project #: TUBITAK 108O582. 2012; 186.

[36] Kaman, H. Yield response of maize genotypes to conventional deficit irrigation and partial root drying (Ph.D. Dissertation). Adana, Turkey: Çukurova University (in Turkish); 2007.

[37] Dinc U, Sarı M, Senol S, Kapur S, Sayın M, Derici MR, Cavusgil V, Gök M, Aydın M, Ekinci H, Ağca N, Schlinchting E. Soils of Cukurova Region, Cukurova Üniversitesi Ziraat Fakültesi Yardımcı Ders Kitabı, No: 26, 2. Baskı. C̦ukurova University Publishing, Adana, Turkey (in Turkish); 1988.

[38] Akgul MA. Modelling water and nitrate budget in left bank irrigation of lower Seyhan plain. Adana, Turkey: Cukurova University, Institute of Natural and Applied Sciences, (in Turkish with English abstract); 2015.

[39] Arnold JG, Kiniry JR, Srinivasan R, Williams JR, Haney EB, Neitsch SL. SWAT and Water Assessment Tool, Input/Output Documentation, Chapter 20 SWAT Input Data: MGT, 243 p. http://swat.tamu.edu/documentation/2012-io. 2012

[40] Tomer MD, Dosskey MG, Burkart MR, James DE, Helmers MJ, Eisenhauer DE. Methods to prioritize placement of riparian buffers for improved water quality. Agroforestry Systems. 2009; 75: 17–25. DOI: 10.1007/s10457-008-9134-5

[41] Gassman PW, Reyes MR, Green CH, Arnold JG. The soil and water assessment tool: historical development, applications, and future directions. Transactions of the ASABE. 2007; 50 (4): 1211–1250.

[42] Jain KS, Tyagi J, Singh V. Simulation of runoff and sediment yield for a Himalayan watershed using SWAT model. Journal of Water Resource and Protection. 2010; 2: 267–281. DOI:10.4236/jwarp.2010.23031

[43] Maharjan M, Babel MS, Maskey S. Reducing the basin vulnerability by land management practices under past and future climate: a case study of the Nam Ou River Basin, Lao PDR. Hydrology and Earth System Sciences Discuss. 2014; 11: 9863–9905. DOI:10.5194/hessd-11-9863-2014.

[44] Li L, Jiang D, Hou X, Li J. Simulated runoff responses to land use in the middle and upstream reaches of Taoerhe River basin, Northeast China, in wet, average and dry years. Hydrological Processes. 2013; 27: 3484–3494. DOI: 10.1002/hyp.9481.

[45] Neitsch SL, Arnold JG, Kiniry JR, Williams JR. SWAT and Water Assessment Tool Theoretical Documentation Version. http://swat.tamu.edu/documentation 2009.

[46] Clayton S, Muleta M. Approaches to control nitrate pollution in the San Joaquin watershed. In: World Environmental and Water Resources Congress Proceedings. 2012, 3: 2232–2235. DOI: 10.1061/9780784412312.224

[47] Akhavan S, Abedi-Koupai J, Mousavi SF, Afyuni M, Eslamian SS, Abbaspour KC. Application of SWAT model to investigate nitrate leaching in Hamadan–Bahar Watershed, Iran. Agriculture, Ecosystems and Environment. 2010; 139: 675–688.

[48] Santhi C, Arnold JG, Williams JR, Dugas WA, Srinivasan R, Hauck LM. Validation of the SWAT model on a large river basin with point and nonpoint sources. Journal of American Water Resources Association. 2001; 37: 1169–1188.

[49] Moriasi DN, Gowda PH, Arnold JG, Mulla DJ, Ale S, Steiner JL. Modeling the impact of nitrogen fertilizer application and tile drain configuration on nitrate leaching using SWAT. Agriculture Water Management. 2013; 130: 36–43. DOI: 10.1016/j.agwat.2013.08.003

[50] Pintar M, Kompare B, Uršič M, Bremec U, Gabrijelčič E, Sluga G, Globevnik L. The impact of land use on nutrient concentration in upper streams of waters in Slovenia. Integrated Watershed Management: Perspectives and Problems. Springer, Netherlands; 2010. pp. 190–199. DOI: 10.1007/978-90-481-3769-5_16

[51] Neitsch SL, Arnold J, Kiniry JR, Williams JR. Soil and Water Assessment Tool (SWAT)–Theoretical Documentation, USDA Agricultural Research Service, Temple, Texas, 2001.

[52] Cau P, Paniconi C. Assessment of alternative land management practices using hydrolog-ical simulation and a decision support tool: Arborea agricultural region, Sardinia. Hydrology and Earth System Sciences Discussion. 2007; 11: 1811–1823.

[53] Pongpetch N, Suwanwaree P, Yossapol C, Dasananda S, Kongjun T. Using SWAT to assess the critical areas and nonpoint source pollution reduction best management practices in Lam Takong River Basin, Thailand. Environment Asia 8.1. 2015; 41–52. DOI: 10.14456/ea.2015.6

[54] Ullrich A, Volk M. Application of the soil and water assessment tool (SWAT) to predict the impact of alternative management practices on water quality and quantity. Agricul-tural Water Management 2009; 96 (8): 1207–1217.

[55] Abraham LZ, Roehrig J, Chekol DA. Calibration and validation of SWAT hydrologic model for meki watershed, Ethiopia. In: Conference on International Agricultural Research for Development, Tropentag 2007 University of Kassel-Witzenhausen and Uni-versity of Göttingen; 9–11 October 2007.

[56] Abbaspour KC, Yang J, Maximov I, Siber R, Bogner K, Mieleitner J, Zobrist J, Srinivasan R. Spatially-distributed modelling of hydrology and water quality in the prealpine/alpine Thur watershed using SWAT. Journal of Hydrology. 2007; 333: 413–430.

[57] Glavan M, Pintar M, Urbanc J. Spatial variation of crop rotations and their impacts on provisioning ecosystem services on the river Drava alluvial plain. Sustainability of Water Quality and Ecology. 2015; 5: 31–48.

[58] Benaman J, Shoemaker CA, Haith DA. Calibration and validation of Soil and water assessment tool on an agricultural watershed in upstate New York. Journal of Hydrology Engineering. 2005; 10 (5): 363–374.

[59] Moriasi DN, Arnold JG, Van Liew MW, Bingner RL, Harmel RD, Veith TL. Model evaluation guidelines for systematic quantification of accuracy in watershed simulations, Transactions of the ASABE. 2007; 50 (3): 885–900.

[60] Allen RG, Pereira LS, Raes D, Smith M. Crop evapotranspiration: guidelines for computing crop water requirements. Rome: FAO, (Irrigation and Drainage Paper, 56); 1998. 300 p.

[61] Narasimhan B, Srinivasan R, Arnold JG, Di Luzio M. Estimation of long-term soil mois-ture using a distributed parameter hydrologic model and verification using remotely sensed data. Transactions of the ASAE. 2005; 48(3): 1101–1113.

[62] Ibrikci H, Cetin M, Karnez E, Wolfgang AF, Tilkici B, Bulbul Y, Ryan J. Irrigation-induced nitrate losses assessed in a Mediterranean irrigation district. Agricultural Water Manage-ment. 2015; 148: 223–231.

[63] Ulger A, Ibrikci H, Cakir B, Guzel N. Influence of nitrogen rates and row spacing on corn yield, protein content, and other plant parameters. Journal of Plant Nutrition. 1997; 12: 1697–1709.

7

Metals Toxic Effects in Aquatic Ecosystems: Modulators of Water Quality

Stefania Gheorghe, Catalina Stoica,

Gabriela Geanina Vasile, Mihai Nita-Lazar,

Elena Stanescu and Irina Eugenia Lucaciu

Additional information is available at the end of the chapter

Abstract

The topic of this work was based on the assessment of aquatic systems quality related to the persistent metal pollution. The use of aquatic organisms as bioindicators of metal pollution allowed the obtaining of valuable information about the acute and chronic toxicity on common Romanian aquatic species and the estimation of the environment quality. Laboratory toxicity results showed that Cd, As, Cu, Zn, Pb, Ni, Zr, and Ti have toxic to very toxic effects on *Cyprinus carpio*, and this observation could raise concerns because of its importance as a fishery resource. The benthic invertebrates' analysis showed that bioaccumulation level depends on species, type of metals, and sampling sites. The metal analysis from the shells of three mollusk species showed that the metals involved in the metabolic processes (Fe, Mn, Zn, Cu, and Mg) were more accumulated than the toxic ones (Pb, Cd). The bioaccumulation factors of metals in benthic invertebrates were subunitary, which indicated a slow bioaccumulation process in the studied aquatic ecosystems. The preliminary aquatic risk assessment of Ni, Cd, Cr, Cu, Pb, As, and Zn on *C. carpio* revealed insignificant to moderate risk considering the measured environmental concentrations, acute and long-term effects and environmental compartment.

Keywords: metals, fish, crustaceans, benthic invertebrates, toxicity, LC50, MATC, bioaccumulation, risk

1. Introduction

Metals are constantly released in aquatic systems from natural and anthropic sources such as industrial and domestic sewage discharges, mining, farming, electronic waste, anthropic accidents, navigation traffic as well as climate change events like floods (**Figure 1**) [1, 2]. Moreover, metals are easily dissolved in water and are subsequently absorbed by aquatic organisms such as fish and invertebrates inducing a wide range of biological effects, from being essential for living organisms to being lethal, respectively. In spite of the fact that some metals are essential at low concentrations for living organisms, such as (i) micronutrients (Cu, Zn, Fe, Mn, Co, Mo, Cr, and Se) and (ii) macronutrients (Ca, Mg, Na, P, and S); at higher concentrations, they could induce toxic effects disturbing organisms' growth, metabolism, or reproduction with consequences to the entire trophic chain, including on humans [3]. In addition, the non-essential metals such as Pb, Cd, Ni, As, and Hg enhance the overall toxic effect on organisms even at very low concentrations. High levels of metals in the environment could be a hazard for functions of natural ecosystems and human health, due to their toxic effects, long persistence, bioaccumulative proprieties, and biomagnification in the food chain [4, 5]. In this context, metal pollution is a global problem; therefore, the international regulations demanded for water quality compliance with the quality standards both in surface water or groundwater and in biota [6–9]. Currently, in accordance with the European Water Framework Directive (EU-WFD, 2000/60/EC), the ecological status of water bodies is assessed based on five biological indicators such as phytoplankton, macrophytes, phytobenthos, benthic invertebrates, and fish alongside with chemical and hydromorphological quality elements. Due to the fact that biota has the ability to accumulate various chemicals, it has been extensively used to measure the effects of metals on aquatic organisms as an essential indicator of water quality [10]. The mollusks [11–14] and fish [3, 15] are the most used organisms as bioindicators of metal pollution in water or sediment.

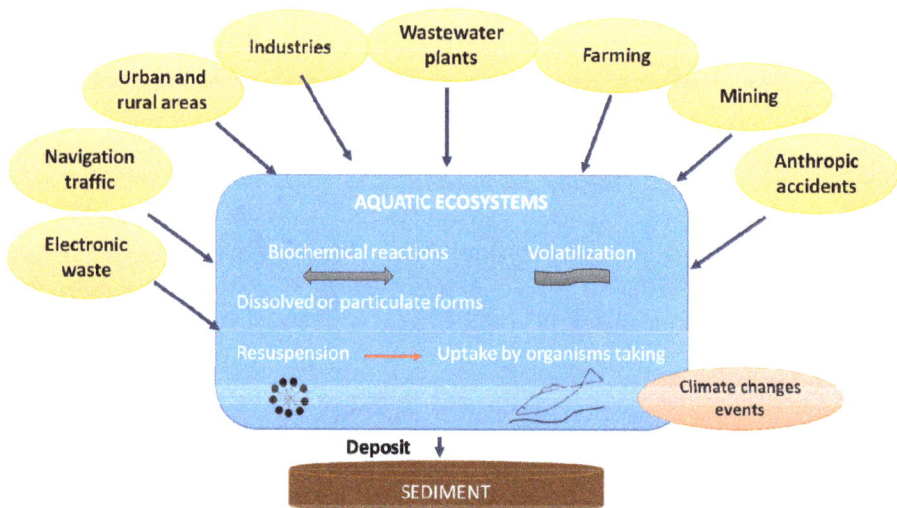

Figure 1. Sources of metal contamination affecting aquatic ecosystems.

The proposed topic of this chapter is based on the assessment of aquatic systems quality linked to persistent metal pollution. The chapter includes an extensive literature review concerning the impact of heavy metals on aquatic systems followed by an experimental part based on metal distribution and toxicity effect on the Romanian surface waters. Due to the European economic and strategic importance of Danube Delta, the final receptor of Danube's flow, the toxic effect of various metal concentrations (Ni, Zn, Cu, Cd, As, Cr, Pb, Co, Ti, Zr, Fe, Mn, etc.) was analyzed.

2. Metal ecotoxicology

2.1. Heavy metal bioavailability to aquatic organisms

Unlike organic chemicals, the majority of metals cannot be easily metabolized into less toxic compounds, a characteristic of them being the lack of biodegradability. Once introduced into the aquatic environment, metals are redistributed throughout the water column, accumulated in sediments or consumed by biota [16]. Due to desorption and remobilization processes of metals, the sediments constitute a long-term source of contamination to the food chain. Metal residues in contaminated habitats have the ability to bioaccumulate in aquatic ecosystems—aquatic flora and fauna [17], which, in turn, may enter into human food chain and result in health problems [18]. Metal accumulation in sediments occurs through processes of precipitation of certain compounds, binding fine solid particles, association with organic molecules, co-precipitation with Fe or Mn oxides or species bounded as carbonates—according to the physical and chemical conditions existing between the sediment and the associated water column [19, 20].

Metal bioavailability is defined as the fraction of the total concentration of metal which has the potential to accumulate in the body. The factors that control the bioavailability of metals (**Figure 2**) are the following: the organism biology (metals assimilation efficiency, feeding strategies, size or age, reproductive stage); metal geochemistry (distribution in water—sediment, suspended matters, and metal speciation) [21, 22]; physical and chemical factors (temperature, salinity, pH, ionic strength, concentration of dissolved organic carbon, total suspended solids) [23, 24].

Metal bioavailability controls their accumulation in aquatic organisms. The metals uptaken paths are through the permeable epidermis if metals are in dissolved forms or through the food ingestion if metals are in particulate forms. Metal speciation, the presence of organic or inorganic complexes, pH, temperature, salinity, and redox conditions [24] are the main factors that could modulate metal toxicity. The ingestion uptake depends on similar factors, plus the rate of feeding, intestinal transit time, and the digestion efficiency [25].

Many studies have shown that the free hydrated metallic ion is the most bioavailable form for Cu, Cd, Zn [26], and Pb [27], but some exceptions have been reported [28]. Thus, the importance of other chemical forms of dissolved metals and complexes formed with suitable organic ligands with low molecular weight should not be neglected. It has been found that the presence of organic binders increases the bioavailability of Cd in mussels and fish, by facilitating the diffusion of the hydrophobic compound in the lipid membrane. The organic compounds of metals could be more bioavailable than the ionic forms [29]. For instance, the organic mercurial compounds are lipid-soluble and penetrate quickly the lipid membranes, increasing the toxicity compared to mercuric chloride which is not lipid-soluble [30].

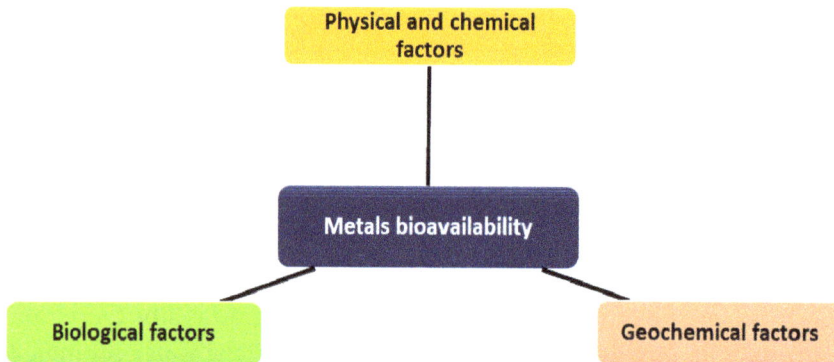

Figure 2. The main control factors that influence metal bioavailability.

The adsorption on suspended solids affects the total concentration of metals present in water. The association between solid particles and metals is also critical for the metal uptake into organisms through food ingestion [31]. The suspended solids accumulate the insoluble metal compounds, but under certain conditions, the metal reached the interstitial water being dissolved. Heavy metal concentrations from sediments or suspended solids are much higher than in water, so a small fraction of them could be an important source for bioaccumulation in planktonic and benthic organisms [32]. The dynamics of different forms of metals in the aquatic environment is not fully understood, so new studies are required to analyze the different accumulation/bioaccumulation pathways based on dissolved or suspended metal forms.

Other studies highlighted that bioavailability of metals in bivalve mollusks depends on sediment particle size due to their filter feeding behavior. If the particles were coated with bacterial extracellular polymers or fulvic acids, the Cd, Zn, and Ag bioavailability was significantly increased. Overall, the binding of metal decreased the bioavailability of metals from the sediment [28, 33].

2.2. Metal toxicity effects

After metal ingestion, this is specifically transported by lipoproteins into different body compartments (organs, blood, or other physiological structures) where they can be specifically oriented to different centers: (i) *action centers* where the toxic metal interacts with an endogen macromolecule (protein or ADN) or a certain cellular structure inducing toxic effects for all body; (ii) *metabolism centers* where the metal is processed by detoxified enzymes; (iii) *storage centers* where the metals are collected in a toxic inactive state; and (iv) *excretion centers* where the metals are disposed.

The heavy metal overload has inhibitory effects on the development of aquatic organisms (phytoplankton, zooplankton, and fish) [34, 35]. The metallic compounds could disturb the oxygen level and mollusks development, byssus formation, as well as reproductive processes. Several histological changes such as gill necrosis or fatty degeneration of the liver occur in the fish and crustaceans [36, 37]. Assessments at the cellular level enable to understand the action of toxic metals on the enzymatic metabolism and physiology of the aquatic organisms.

The lethal effects of metals in crustaceans were induced by the inhibition of enzymes involved in cellular respiration. The histological changes observed in fish and crustaceans after chronic exposure to metals are the result of antioxidant enzymes inhibition [38–41]. The effects on organisms' growth and development were triggered by the inhibition of enzymatic systems involved in protein synthesis and cell division. The metal type modulates the bioaccumulation level and enzymatic systems vulnerability generating a multitude of effects, toxic or not [42, 43].

In order to understand the interaction mechanism between the toxic metals and the aquatic organisms and how organisms answer to metal contamination, more information on bioavailability is needed [44].

At the present, many studies on the assessment of acute and chronic toxicity of metals mentioned the following parameters: survival, growth, development, reproduction, behavior, accumulation, effects on enzyme systems, etc. In **Table 1**, the values of acute (LC50) and chronic (MATC/NOEC/LOEC) toxic concentration for fish and planktonic crustaceans according to PAN Pesticide Database—Chemical Toxicity Studies on Aquatic Organisms [45] are exemplified. The studies highlighted that the toxic concentration intervals depend on the species, exposure time, age of specimens, type of toxicity test type, and laboratory conditions.

Metal	Fish (*Cyprinus carpio*)		Crustaceans (*Daphnia magna*) EC50
	LC50 (96 h)	MATC/NOEC/LOEC	(48 h)
Ni	1.3–10.4 mg/L	NOEC 50 µg/L LOEC 3.50; 13 µg/L; 1.97 mg/L	1 g/L
Zn	0.45–30 mg/L	NOEC 2.60 µg/L; 0.43; 4.20 mg/L	0.35–3.29 mg/L
Cd	2–240 µg/L 3–17.05 mg/L	NOEC 0.02–37 µg/L LOEC 6.7–440 µg/L	24–355.4 µg/L
As	0.49 mg/L (*Carassius sp.*) 0.9 mg/L (*Pimephales sp.*)	LOEC 25 µg/L	3.8 mg/L
Cr	14.3–93 mg/L	NOEC 0.19–17 µg/L LOEC 25 µg/L	22–160 µg/L
Pb	0.44–2 mg/L	NOEC 0. 07; 128 µg/L LOEC 0.03–128 µg/L	4.4; 5.7 mg/L (*Daphnia sp.*)
Sb	6.2–8.3 mg/L (*Cypridon sp.*)	NOEC 6.2 mg/L	>1 g/L
Mn	0.1–15.61 mg/L (*Oncorhynchus sp.*)	–	40 mg/L

LC (EC) 50—lethal concentrations for 50% of tested organisms after 96 or 48 h; MATC—maximum acceptable toxicant concentration in aquatic systems; NOEC—no observed effect concentration; LOEC—low observed effect concentration.

Table 1. Literature acute and chronic toxicity values [45–47].

2.3. Metal bioaccumulation and bioamplification

Monitoring the toxicity and accumulation of metals into the aquatic biota or sediment is mainly performed for assessing both the surface water quality and ensure food safety, respectively, as well as for the compliance with the directives. The toxicity and accumulation parameters are used for various environment monitoring programs such as wastewater discharges or various risk assessments of natural and anthropogenic events (floods or dredging activities) and also for identifying the source of metal contamination [48, 49].

According to the European document COM (2011)—876 final—2011/0429 (COD) (2012/C 229/22) amending the Directive 2000/60/EC and 2008/105/EC on priority substances in the field of water policy, new concentration limits of a number of harmful chemical compounds were allowed for the aquatic biota (fish, mollusks, or crustaceans). For instance, diphenyl brominated, fluoranthene, hexachlorobenzene, hexachlorobutadiene, benzene compounds, dicofol, perfluorooctane sulfonic acid and its derivatives, dioxins and dioxin-type compounds, cyclohexa-bromo-dodecane, heptachlor epoxide, and heptachlor have a concentration values in the range 6.7×10^{-3} to 167 mg/kg wet weight. The rest of the chemical compounds were not yet amended, which represents a considerable research opportunity to assess their chronic toxicity, bioaccumulation, and subsequently to set their maximum permissible concentration limits in aquatic organisms. Furthermore, the Directive 2008/105/EC of Environmental Quality Standards (EQS) entail values for various chemicals in biota.

More and more studies of various organic and inorganic chemical bioaccumulation/bioconcentration in freshwater organisms revealed induced harmful effects, especially of heavy metals (Hg, As, Cd, Zn, Fe, Pb, Fe, Mn, etc.) [10, 50, 51] and metal nanoparticles [52]. Bioaccumulation remains to be an ongoing highly debated subject. According to United States Geological Survey (USGS) Toxic Substances Hydrology Program, the bioaccumulation represents *"the biological sequestering of a substance at a higher concentration than that at which it occurs in the surrounding environment or medium."* Pollutants can be uptaken by organism directly from the environment or through ingestion of particles [53], and the accumulation occurs when an organism absorbs toxic chemical with a rate faster than the chemical is metabolized. On the contrary, the bioconcentration refers to the chemical uptake from the water only, which could be assessed in the laboratory conditions. The value of the concentration factor index gives information if there is a bioaccumulation (the concentration factor of <1) or a bioconcentration (concentration factor >1) [54]. The understanding of the bioaccumulation process is important because persistent pollutants (such as metals) could increase the toxic potential risk by bioaccumulation in the ecosystem, triggering a long-term effect on the ecosystem which cannot be assessed by laboratory toxicity tests [54]. It is considered that a high bioaccumulation potential does not necessarily imply a high potential for toxicity, and as a result, the toxic effects should be estimated separately. In addition, it was made a distinction between accumulation in a small concentrations range, which occurs due to physiological needs (e.g., Zn) and apparently uncontrolled accumulation (e.g., Cd) [55].

It was observed that the mollusks from the Black Sea have shown a great tendency to accumulate in high concentration Cd and Cu from sediments as well as Cd, Ni, and Cu from water. Data showed that the highest concentrations of heavy metals were found in the digestive tract of fish [56]. Also, the fish *Cyprinus carpio* could differentially bioaccumulate metals inform one organ to another: Zn > Cr > Pb > Cu in muscle; Pb > Cr > Zn > Cu in gills; Pb > Cr > Zn in liver [10]. Moreover, for the same species, it has been shown that gills and liver or kidney were accumulating the following metals: Pb > Cd > Cr > Ni and Pb > Cd > Ni > Cr. On the other hand, bioaccumulation of Pb and Cd was significant in all *C. carpio* tissues [57].

Metal transfer in the aquatic food chain is another interesting environmental topic for many reasons such as the accumulation of metals in aquatic organisms that could transfer up to humans, leading to a potential risk of public health through consumption of contaminated fish [58, 59]. It is known that aquatic organisms can be exposed to high or low concentrations of metals as a result of continuous or accidental release, causing long-term effects. The main uptake pathways of metals in aquatic organisms are direct through the food or sediment particles ingestion and water via epidermis and gills then they are transported inside the cells through biological membranes and ionic channels [60]. Bioconcentration and bioaccumulation of metals into the trophic chain occur if metals are excreted into the water or the contaminated organisms are food for some predator's organisms [61, 62].

The study named "*Ecotoxicology of heavy metals in the Danube meadow*" [55] revealed that the amplification of metal concentrations in the food chains of ecosystems depends on the type of metal and the food chain. The metal accumulation in plants depends on the species, metal type, and ecosystem, especially for species which predominantly take metals from soil/sediment. Benthic gastropods tend to differentially accumulate the metals. The populations which are using the seston as energy source concentrate in many cases metals from different sources: Bivalves shell accumulate Pb, the tissue—Mn, Zn, Cd, and sometimes the amphibians in young stages accumulate Cd, Mn, and Zn.

Metal concentrations such as Fe, Mn, Cu, Cr, and Pb were not amplified in the food chain (benthic fauna-fish-birds), but they were amplified for Zn and Cd. Concentrations of metals were greater at the end of the trophic chains, as follows: vegetation/detritus—terrestrial invertebrates phytophase/detritophage—terrestrial invertebrates' predators—amphibians (Cd, Cr, Pb, and Cu in case of detritus chain and Zn in case of vegetation). The fish always accumulate metals, with some exceptions in the case of Cd, for which the transfer coefficient indicates accumulation in muscle and liver. The transfer of metals from benthic invertebrates to omnivorous fish revealed concentration of Zn and Cu in the liver and Zn in muscle. The Mn, Cr, and Cd metals transfer from omnivorous fish (muscle) to predatory fish, more specifically in their muscles and liver. At the end, the birds that are using contaminated fish as food source will accumulate Fe, Mn, Zn, Cu, and Cd in muscle and all metals (except Cr) in the liver [55].

3. Experimental part

3.1. Occurrence of metals in surface water and sediments

Several monitoring studies performed by INCD ECOIND Bucharest researchers during 2003–2013 in the Danube Delta—Sfantu Gheorghe Branch (sampling points: Mahmudia,

Murighiol, and Uzlina) emphasized some heavy metal concentration patterns in the study area (**Table 2**). The metal concentrations in water were within the limits of Romanian legislations, for class I and class II quality (according to the EU-WFD and the requirements set by the Romanian Law 310/2004 which amends the Law 17/1996). Cu and Ni showed the highest total concentration (**Table 2**, marked lines) among the determined metals.

Metal	Mahmudia (2009–2013)				Murighiol (2003–2013)				Uzlina (2003–2013)			
	Min	Max	Average	SD	Min	Max	Average	SD	Min	Max	Average	SD
Ni	<1.0	24.0	**4.02**	5.79	<1.00	68.1	**12.4**	18.6	<1.00	10.3	**2.60**	2.85
Fe	<20	880	**300**	270	112	3400	**710**	750	80.0	1040	**350**	280
Mn	<2.0	30.0	**10.0**	10.0	3.00	290	**70.0**	80.0	5.00	50.0	**20.0**	10.0
Cd	0.40	0.40	**0.40**	0.00	<0.10	0.50	**0.36**	0.14	<0.10	0.50	**0.37**	0.13
Cr	<0.5	6.00	**2.09**	1.74	<0.50	21.0	**5.38**	6.00	<0.50	21.0	**3.52**	5.33
Cu	2.50	10.5	**5.74**	2.67	0.012	55.3	**12.9**	17.8	0.03	123	**14.5**	26.7
Pb	<2.0	3.20	**2.08**	0.31	<2.00	5.00	**2.15**	1.29	<2.00	5.00	**2.17**	1.27
As	<2.0	2.20	**2.01**	0.05	<2.00	3.90	**1.82**	0.88	<2.00	2.64	**1.73**	0.59
Hg	<0.1	0.24	**0.32**	0.06	<0.10	0.77	**0.15**	0.20	<0.10	0.14	**0.22**	0.10
Zn	<2.0	24.7	**10.1**	7.07	<2.00	56.0	**9.58**	11.4	<2.00	57.0	**8.15**	11.7
Co	<0.5	1.30	**0.66**	0.33	<0.50	5.00	**1.17**	1.65	<0.50	5.00	**1.18**	1.65

Min—the minimum detected concentration; Max—the maximum detected concentration; SD—standard deviation.

Table 2. Occurrence of metals in water of Danube Delta—Sf. Gheorghe (2003–2013) in μg/L [63, 64].

The studies revealed that metals Cu, Pb, Zn, Cr, Ni, Cd, Mn, and Fe were the most abundant in the sediments of the Danube Delta—Sf. Gheorghe Branch sampling sites. The concentrations of these metals ranged with the sampling location and seasonal or natural events, as follows: Cu 4.65–194 mg/kg d.m (dry matter), Pb 4.76–51.3 mg/kg d.m., Zn 17.7–218 mg/kg d.m., Cr 7.5–61.9 mg/kg d.m., Ni 10.8–111 mg/kg d.m., Cd <0.01–1.5 mg/kg d.m. (**Figure 3**). The average value in the period 2009–2013 for Mn was 614.03 mg/kg d.m. and for Fe, it was 20 987 mg/kg d.m. [64, 65].

Alongside metal concentration, several chemical (nutrients, oxygen and pH regime, pesticides, petroleum products, polychlorinated biphenyls) and biological (phytoplankton, zooplankton, and benthic macroinvertebrates) elements were investigated, showing that the organochlorine pesticides and petroleum products exceeded the maximum allowed limits [63, 66].

In addition, other studies performed along Romanian rivers showed that the mining activities had a great impact on sediment ecosystems due to metal pollution. For instance, a study performed during 2003 in Baia Mare (in North Vest of Romania) mining area after a pollution accident showed a high content of heavy metals in Someș River sediment (Cu 104–339 mg/kg, Pb 59–465 mg/kg, Zn 56–2060 mg/kg, Cd 0.05–14.14 mg/kg, CN 0.33–15.86 mg/kg). The detected concentration affected the aquatic ecosystem where the microalgae species disappeared and the number of fish species decreased dramatically compared to the period before the incident. Also many species of mollusks disappeared because their capacity to accumulate large amount of heavy metals was exceeded [67]. In addition, in Rosia Montana area (in the West part of Romania), significant water contamination with heavy metals occurred due to the mining acidic waters from area on two water courses: Rosia and Corna stream. The results showed exceedances of Cu, Cd, Fe, Ni, and Cr, in particular in the Rosia Montana water stream [68]. Along Jiu River (in south of Romania) sediments, heavy metal pollution in most sampling points was recorded according to the pollution load index (PLI) [69].

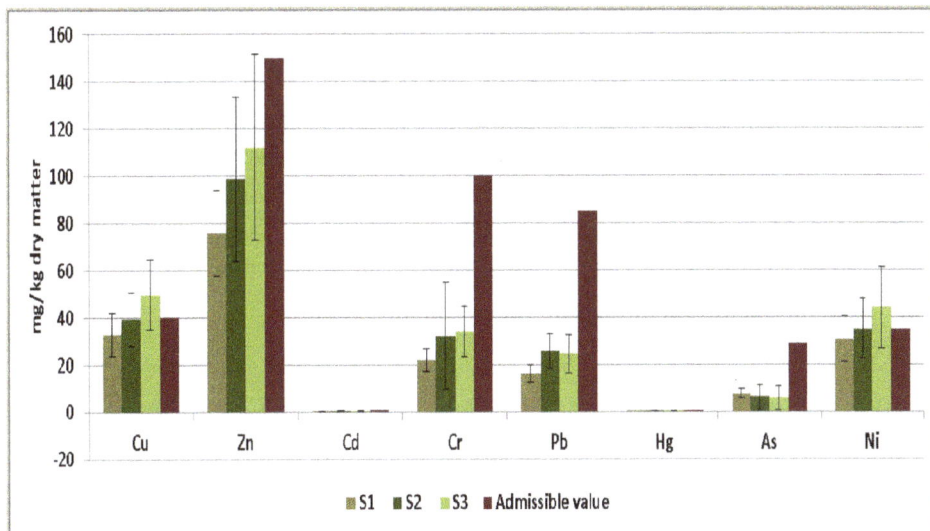

Figure 3. Occurrence of metals in sediments of Danube Delta—Sf. Gheorghe (2003–2013) (average values). S1—Mahmudia, S2—Murighiol, and S3—Uzlina [65].

In the above context, in the following sections will be presented some data concerning the metals effects on freshwater organisms (fish, planktonic crustacean, and mollusks), obtained through laboratory testing or by biological samples collected from contaminated fields.

3.2. Laboratory tests: acute and chronic effects of metals

3.2.1. Materials and methods

The assessment of metals acute and chronic effects was based on fish (*C. carpio*) and planktonic crustacean (*Daphnia magna*) laboratory data. The tested organisms were those recommended by the international ecotoxicology protocols (OECD or ISO), and they are frequently found in Romanian surface waters, easily to acclimatize in laboratory and sensitive to various contaminants. *C. carpio* are in particular the most affected organisms due to the fact they ingest both planktonic and benthic organisms, respectively, and thus, they especially accumulate the contamination from water and sediment. The tests were performed on the following metals: Ni, Zn, Cu, Cd, As, Cr, Pb, Sb, Mn, Ti, and Zr, which were usually detected in the aquatic systems.

3.2.1.1. Sample preparation

For stock solution preparation, a known quantity of metals test as $NiSO_4$, $ZnSO_4$, $CuSO_4$, $CdCl_2/CdSO_4$, As_2O_3, $K_2Cr_2O_7$, $Pb(NO_3)_2$, $SbCl_5$, $MnCl_2x4H_2O$, TiO_2, $ZrCl_4$ was dissolved into the specified volume of dilution water or growth medium. No added solvents have been used, and all substances have been tested under their maximum solubility. The solutions were stirred for 24 h, in the dark at 25°C. The testing solutions were prepared by mixing the appropriate volumes of stock solution with dilution water or growth medium in order to obtain the final concentrations used for testing. Finally, the pH values of tested solutions were situated between 6.5 and 8.5 units.

3.2.1.2. Fish toxicity test procedure

Using OECD methodologies for acute toxicity, the lethal concentrations for 50% of tested organisms were estimated. Metals' long-term toxicities on fish were conducted using an in-house methodology based on the changes in some physiological indicators such as growth rate, mortality, biomass, production, food use and biochemical indicators, hepatic enzyme activity, respectively. **Table 3** presents the technical parameters of fish toxicity tests.

3.2.1.3. Crustacean toxicity test procedure

The toxicity test determined the metal concentration that immobilizes or kills 50% (LC50) of *D. magna* crustacean, after chemical exposure at 20°C ± 2°C in the dark for 24 or 48 h. The test procedure was performed according to OECD 202 using the microbiotest Daphoxkit F Magna provided by MicroBioTests Inc., Belgium. Briefly, the test was performed in three replicates, in multiwall test plates (six rinsing wells and 24 wells for toxicant dilutions) using 20 organisms per each concentration (at least five different concentrations for each metal) and control (untreated standard freshwater). The mortality/immobility percentage of organisms was registered after 24 and 48 h.

3.2.1.4. Data processing and statistics

The acute effect concentration values in the fish and crustacean tests were calculated using probity analysis method, based on exponential regression relationship between cumulative percentages of mortality (expressed as probity units) for each exposure period against logarithmic concentrations of test substance. For each result, standard deviations were calculated.

3.2.2. Results and discussion

3.2.2.1. Fish toxicity

Acute toxicity tests provide a measure of toxicity for a target species under specific environmental situations and could suggest a rapid and severe effect of contaminants. Acute and chronic toxicity test mimicked the metals accidental release or long-term accumulation in sediment [67–70]. The carp fish LC50-96h values showed different responses in direct correlation with the metals type and concentration. The LC50-96h values were 0.16, 0.28, 0.31, and 0.40 mg/L for Cd, Ti, Zr, and As (**Figure 4**), 2.17, 12.2, 30.10, and 65.8 mg/L for Cu, Zn, Pb, and Ni (**Figure 5**), 120, and 758 mg/L for Cr and Sb, obtained from two replicates for each metal (**Table 4**).

Test conditions	OECD 203 (acute tests)	In-house procedure (chronic tests)
Holding of fish	Acclimatization of fish in laboratory tanks for 3 weeks	
Limit test	One concentration selected according to scientific literature	MATC estimated = LC50-96 h × 0.1
Definitive test		
Test concentrations in definitive test	Five concentrations in a geometric series	Two concentrations (under or over the estimated MATC)
Type of test	Static	Discontinuous (renewal solutions at 24 h)
Time of exposure	96 h	60 days
Fish species and characteristics	*Cyprinus carpio* (1 year) 10 exemplars/test solution, 5–7 cm, 10–15 g/exemplary	*Cyprinus carpio* (2 years) 20–30 exemplars/test solution, 12–14 cm, 25–30 g/exemplary
Fish source	Romanian specialized fish farm	
Testing vessels	10 L	100 L
Temperature, oxygen concentration, pH, light	18–25°C, ≥4 mgO$_2$/L, pH 6.5–8.5 (daily measuring), 12- to 16-h photoperiod daily. Mean of water total hardness 13 mg/L CaCO$_3$	
Feeding	Not food	2% from the surviving lot weight/day
Control test	All toxicity tests were carried out in the same time with a control test	
Replicates	Two replicates/test/metal	
Analytical control in test solutions	Inductively coupled plasma atomic emission spectrometry (ICP-OES)	
Toxicity criteria	Organisms mortalities and visible abnormalities (at 24, 48, 72, and 96 h)	Growth instant rate, mortality rate, biomass mean, production, used food rate, and biochemical indicators—hepatic enzymes activity—GOT and GPT
Results treatment	Probity analysis method based on the exponential regression model between the mortality (probity units) and the log of concentrations of the metal	Comparative analyses with the controls
End points	Lethal concentrations for 50% of tested fish after 96 h of exposure (LC50-96 h)	Maximum acceptable toxicant concentration in aquatic systems (MATC)

Table 3. Test conditions of acute and chronic toxicity tests.

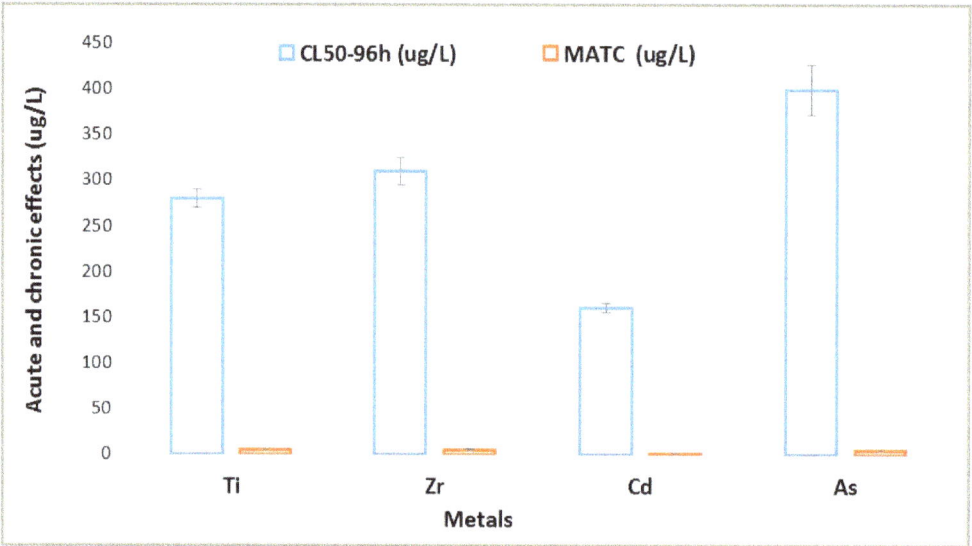

Figure 4. Acute and chronic toxicity of Ti, Zr, Cd, and As classified in very toxic class for *Cyprinus carpio*.

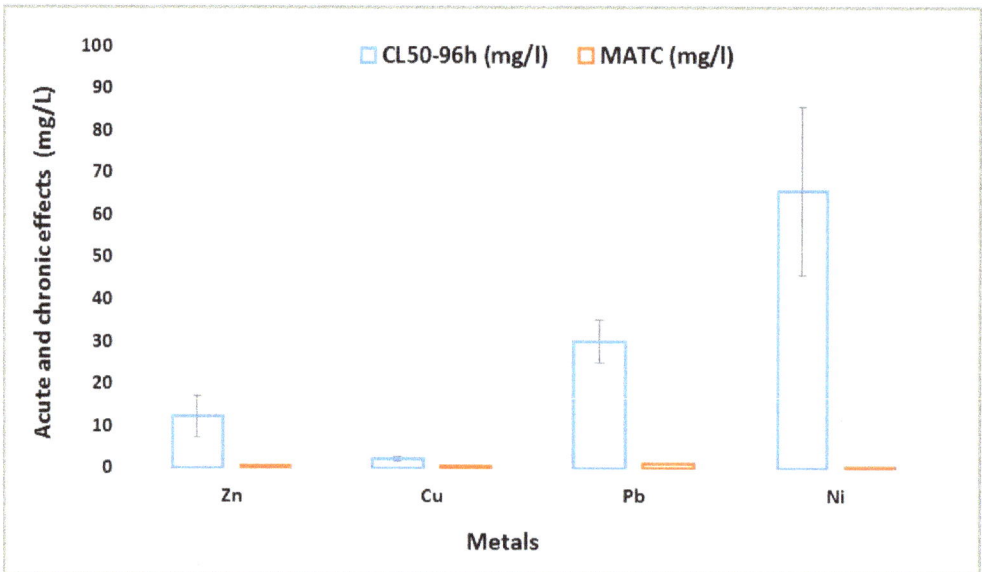

Figure 5. Acute and chronic toxicity of Zn, Cu, Pb, and Ni classified in toxic class for *Cyprinus carpio*.

According to Global Harmonization System for chemical classification and labeling, Cd, Ti, Zr, and As were the most toxic metals for fish. Cd, Ti, Zr, and As showed to be very toxic compared with the other analyzed metals. Research studies revealed similar acute toxicity intervals: 6.16–47.58 mg/L for Ni, 0.15–21.4 mg/L for Zn, 0.28–34.5 mg/L for Cu, 0.005–7.92 mg/ L for Cd and 90 to >139 mg/L for Cr [71]. The maximum acceptable toxicant concentration (MATC) is a value calculated from chronic toxicity tests [72] in order to set water quality norms for aquatic life protection.

Metals	Cyprinus carpio		Daphnia magna	G.D. 351/2005[a]	Directive 105[b] (µg/L)	National plan[c] (µg/L)	Toxicity class[d]
	LC50-96h (mg/L)	MATC (mg/L)	LC50-48h (mg/L)	(µg/L)			
Ti (TiO$_2$)	0.28 ± 0.01	0.005 ± 0.001	5.56 ± 0.8	–	–	–	Very toxic—fish
Zr (ZrCl$_4$)	0.31 ± 0.01	0.005 ± 0.002	91.20 ± 10	–	–	–	Very toxic—fish
Ni (NiSO$_4$)	65.8 ± 20.0	0.10 ± 0.02	–	20	20	4–34	Toxic—fish
Zn (ZnSO$_4$)	12.2 ± 5.0	0.60 ± 0.01	–	5	–	11.80–73	Toxic—fish
Cu (CuSO$_4$)	2.17 ± 0.50	0.05 ± 0.01	–	100	–	1.22–10	Toxic—fish
Cd (CdCl$_2$/CdSO$_4$)	0.16 ± 0.001	0.001 ± 0.0005	0.14 ± 0.01	5	0.2	–	Very toxic—fish and daphnia
As (As$_2$O$_3$)	0.40 ± 0.02	0.005 ± 0.001	–	10	–	49	Very toxic—fish
Cr (K$_2$Cr$_2$O$_7$)	120 ± 22	1.00 ± 0.01	0.81 ± 0.02	50	–	8.8	Very toxic—daphnia
Pb (Pb(NO$_3$)$_2$)	30.1 ± 5.0	1.00 ± 0.02	–	10	7.2	–	Toxic—fish
Sb (SbCl$_5$)	758 ± 24	0.060 ± 0.001	148 ± 21	5	–	–	Non-toxic—fish and daphnia
Mn (MnCl$_2$×4H$_2$O)	>53 ± 8	–	–	–	–	–	Non-toxic—fish

[a] Governmental Decision no. 351/2005 concerning the hazard chemical discharge.
[b] Directive 2008/105/EC on environmental quality standards in the field of water policy.
[c] National Plan of River Basin Management (2016-2021)—Annex 6.1.3B.
[d] According to REACH 1907/2006; Regulation (EC) 1272/2008; Regulation (EU) 286/2011; Global Harmonization System for chemical classification and labeling (GHS) Revision 2011. The toxicity class was decided on the highest toxicity of target organisms.

Table 4. In-house toxicity data of metals for fish and crustacean in relation with the national and international norms for metals limits in surface water.

Experimental exposure of fish for 60 days to different concentrations of metals revealed different long-term effects. The final results showed no effects concentrations on target organisms, assessment of environmentally safe concentrations, respectively. The calculation of MATC values started by multiplication of the LC50-96h of each metal with an application factor of 0.1 (**Table 2**). The monitored physiological parameters from chronic test revealed that Cd is non-toxic at 0.001 mg/L, Ti, Zr, and As were safety to 0.005 mg/L, Cu at 0.05 mg/L, Sb at 0.06 mg/L, Ni at 0.10 mg/L, Zn at 0.60 mg/L, Cr and Pb at 1.00 mg/L, comparative with the controls (**Figures 4** and **5, Table 4**). Similar values for Cu (0.012 mg/L) and Zn (0.5 mg/L) were also obtained in other studies [73].

3.2.2.2. Crustaceans toxicity

Toxicity tests on *D. magna* crustaceans showed various toxicities of metals; the LC50-48h showed 0.14 mg/L for Cd, 0.81 mg/L for Cr, 5.56 mg/L for Ti, 91.2 mg/L for Zr, and 148 mg/L for Sb. Cd and Cr showed the highest toxicities and were classified in very toxic chemicals class for *Daphnia* sp. (**Figure 6**). Similar literature values were reported for Cr between 0.02 and 0.05 mg/L [71] and for Cd between 0.024 and 0.355 mg/L [45].

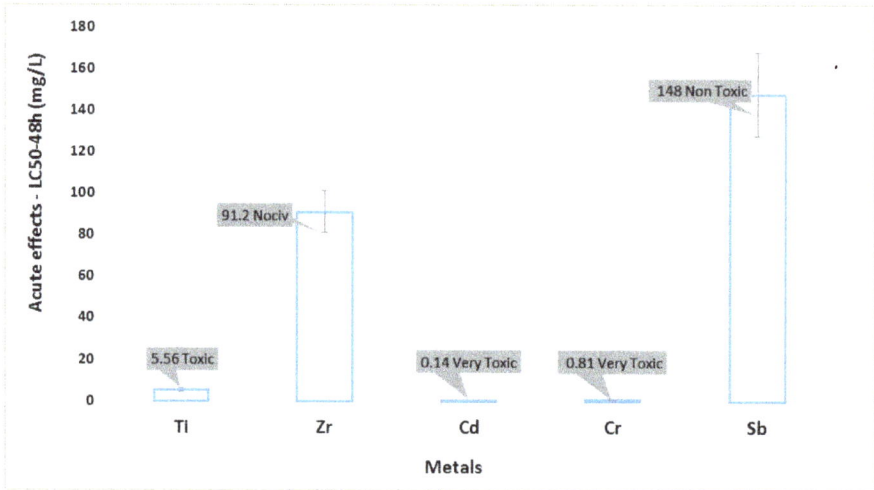

Figure 6. Acute toxicity of Ti, Zr, Cd, Cr, and Sb for *Daphnia magna*.

The surface water quality norms require specific limits only for few very toxic and toxic metals. For example, Ti, Zr, Cd, and Pb norms are not established by the National Plan of River Basin Management—Annex 6.1.3B, despite of their acute toxic effects at very low concentrations (**Table 4**). Also the Directive 2008/105/EC on environmental quality standards in the field of water policy sets limits only for Ni, Cd, and Pb. The present limits assure the protection of aquatic organisms, especially for fish and planktonic crustaceans.

3.3. Field test: bioaccumulation

In order to assess the impact of metals in the field, the following sections present some preliminary data concerning the metal bioaccumulation into benthic invertebrates (mollusks).

3.3.1. Materials and methods

3.3.1.1. Studied area characterization

The studied area was focused on a highly sinuous channel, located on the southeast area of the Danube Delta (Sf. Gheorghe Branch) receiving 22% of Danube's water flow. The Sf. Gheorghe Branch has a width varying between 150 and 550 m, and the water depth varies between 3 and 27 m. The sampling sites location was selected taking into consideration the changes in the Sf. Gheorghe Branch morphology as a result of the pressure from anthropic and environmental factors. Iron Gates I dam construction on Danube River led to a 10% decrease in the suspended sediment amount at Isaccea station. Moreover, the Iron Gates II dam building induced a 50% decrease in suspended sediment at Isaccea. These constructions alongside meander modification (during the years 1984–1988) have produced major changes in sediment distribution. The establishing of space location was performed using GPS type system map 60CSx—Garmin [74].

In addition, the anthropic activities undertaken to strength the banks against coastal erosion led to meanders cutoff, which in turn caused continuous biotope degradation. These changes negatively impacted the ecosystem functions by reducing the structure of the main and constant ecological communities, the benthic invertebrates. So that, to characterize metal bioaccumulation (in benthic invertebrates), two representative sampling sites were selected considering the pressure resulted from anthropic and environmental factors (Murighiol and Uzlina)—**Figure 7**. At temporal scale, this study was conducted during summer and autumn of 2013.

Figure 7. Location of sampling sites in Danube Delta (Sfantu Gheorghe Branch) (St 1—Murighiol; St 2—Uzlina).

3.3.1.2. Sampling collection

The sediment samples for both benthic invertebrates and metal analysis were collected in two replicates using a Van Veen grab, according to the following methodologies: EN ISO 5667-1:2008, ISO SR 5667-6:2009, SR ISO 5667-12:2001 and EN ISO 9391:2000. Surface sample unit was of 255 cm^2, and the sampling depth was of 10 cm. The analysis of benthic invertebrates was performed according to SR EN ISO 8689-1:2003. The species identification was performed using a Motic stereomicroscope. The results were calculated taking into consideration the wet biomass.

3.3.1.3. Sample preparation

The biota samples were dried at 40°C (24 h) and crushed then about three grams of biological sample were dissolved in aqua regia (a mixture of suprapure acids HCl 30 and 65% HNO$_3$ in the report 21–7 mL). The mixture was mineralized using a sand bath until complete dissolution. After cooling, the samples were filtered on paper filter (porosity <45 µm) in a 50-mL volumetric flask and filled with ultrapure water. The metal content in the samples was determined by inductively coupled plasma optical emission spectrometry. A calibration curve in the range of 0.1–0.5 mg/L (As, Se, Sb, Cd, Cr, Cu, Co, Fe, Mn, Mg, Ni, Pb, Zn) was performed using a

Certified Reference Material solution (100 mg/L Multi Element Standard Solution, Certipur, Merck). The quality control of the data was carried out according to Quality Control Standards 21A, 100 mg/L, produced by PerkinElmer. A reagent blank in order to estimate the metal contents from acids was prepared.

The mollusks (two bivalves' species: *Unio pictorum* and *Anodonta cygnea*) and one gastropod species (*Viviparus viviparus*) were selected in this study (**Figure 8**) as they prevail in the total biomass of benthic invertebrate community structure, and they are widely used as bioindicators for water quality. Their shells were subjected to metals detection, because they are formed throughout mollusks life and their chemical composition is an integral index to describe the composition of the aquatic environment over time [75]. Bioaccumulation factors of metals were calculated for each tested species.

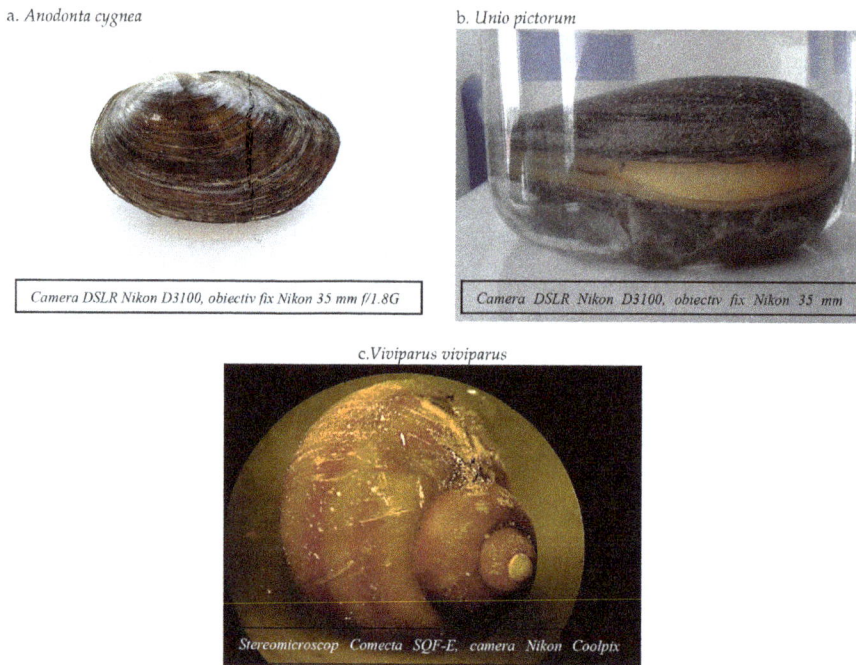

a. *Anodonta cygnea*

b. *Unio pictorum*

Camera DSLR Nikon D3100, obiectiv fix Nikon 35 mm f/1.8G

Camera DSLR Nikon D3100, obiectiv fix Nikon 35 mm

c.*Viviparus viviparus*

Stereomicroscop Comecta SQF-E, camera Nikon Coolpix

Figure 8. The analyzed benthic macroinvertebrates species.

3.3.2. Results and discussion

3.3.2.1. Metal accumulation in benthic organisms

This study included metal analysis results in the bivalves and gastropod shells from Murighiol and Uzlina sampling site. Other researchers [76–79] performed their studies as well using the same biological model, mollusk shells, for the metal accumulation analysis.

The mollusks have the largest representation and are the most valuable groups among the benthic invertebrates' communities due to the fact they are dominant in the total benthic community biomass and represent a basic food for the next trophic level (e.g., fish).

Two types of bivalve species identified at Uzlina and Murighiol were selected for metal analysis, respectively: *U. pictorum, A. cygnea*. Also from Gasteropoda, the species *V. viviparus* were selected (**Figure 8**). For each species, the dry and wet biomasses were determined (**Table 5**).

Sampling point/month	Species	Wet biomass (g)	Dry biomass (g)
Murighiol/July	*Unio pictorum*	25.38	24.49
Uzlina/July	*Viviparus viviparous*	10.96	0.88
	Unio pictorum	39.64	35.44
Uzlina/September	*Anodonta cygnea*	32.73	30.16
	Unio pictorum	19.68	18.58

Table 5. Dry and wet biomass values of the selected species.

The Biota Sediment Accumulation Factor (BSAFsed) was calculated using the equation: $BSAFsed = Cb/Csed$, where Cb is the metal concentration in biota/organism and $Csed$ is the metal concentration in the sediment sample [80].

At Murighiol sampling site, in the *U. pictorum*, shells (collected in July) were recorded the highest values for Cu, Ni, and Zn. Moreover, the Cu concentration in sediment was 47 mg/kg d.m. over the set limit. It was estimated that 4% of the Cu concentration, 2% of Zn, and 1% of Ni were found in *U. pictorum* shell species. The BSAFsed values were <0.05 (**Table 6**). Also, various metals were detected in the *U. pictorum* and *V. viviparous* shells from Uzlina, in July (**Table 7**).

Metal	Cb* *Unio pictorum*	Csed*	Csed*	BSAFsed 2009–2013**
As	<0.05	12.2	9.61	0.004
Cd	<0.01	–	0.50	–
Cu	1.97	47.0	35.1	0.04
Cr	0.12	27.6	31.6	0.004
Co	0.05	9.41	8.37	0.005
Fe	73.2	–	14,895	–
Mn	230	–	464	–
Ni	0.60	35.0	30.8	0.02
Pb	<0.05	25.6	22.3	0.002
Se	0.44	–	–	–
Sb	<0.05	–	–	–
Zn	1.17	91.7	88.5	0.01
Mg	62.5	–	–	–

* Average of metal concentrations (for two replicates) expressed in mg/kg d.m.
** Csed 2009–2013—average of the metal concentration detected in sediment from 2009 to 2013 [63, 65].

Table 6. Metal concentration (mg/kg d.m) in the shell of *Unio pictorum* at Murighiol in July 2013.

Metal	Cb* Unio pictorum	BSAFsed	Cb* Viviparus viviparus	BSAFsed	Csed*	Csed* 2009–2013**
As	<0.05	0.006	<0.05	0.006	7.75	9.30
Cd	<0.01	–	<0.01	–	–	0.51
Cu	2.61	0.05	2.60	0.05	54.7	47.0
Cr	<0.01	0.0003	0.42	0.01	29.6	29.2
Co	0.11	0.01	0.19	0.02	10.8	9.84
Fe	140	–	279	–	–	20987
Mn	58.7	–	30.0	–	–	614
Ni	0.34	0.0085	0.58	0.015	40.0	39.3
Pb	<0.05	0.002	0.15	0.006	26.7	21.3
Se	<0.09	–	<0.09	–	–	–
Sb	<0.05	–	<0.05	–	–	–
Zn	0.90	0.006	3.87	0.02	158	96.9
Mg	154	–	211	–	–	–

* Average of metal concentrations (for two replicates) expressed in mg/kg d.m.
** Csed 2009–2013—average of the metal concentration detected in sediment from 2009 to 2013 [63, 65].

Table 7. Metal concentration (mg/kg d.m) in the shell of *Unio pictorum* and *Viviparus viviparus* at Uzlina in July 2013.

Concentrations of As, Cd, Cr, Fe, Pb, Se, Sb, Mg did not showed significant values in shells of analyzed benthic organisms. The metals Cu, Ni, and Zn were present in sediment over the set limits of national norms inducing their accumulation in shells. The highest values of Cu and Zn were both in *U. pictorum* and *V. viviparus* (**Table 7**). Similar concentrations of Zn, Cu, Pb, Cd, and Co in *V. viviparus* were found in the River Dnieper in the same gastropod shells [81].

Metal concentrations showed a lower magnitude in mollusk shells than in their bodies, and this result could be explained by the fact that metals were accumulated in shell only after they were absorbed by the body. The bioaccumulation selectivity of metals in gastropod shells follows the next order: Fe > Mn > Zn > Cu > Pb > Co > Cd. Thus, the quantitative distribution of metals in mollusk shells is considered by the level of biochemical involvement, metabolic processes, their toxicity degree as well as the bioavailability to aquatic organisms [81].

Some studies [82] revealed that Fe belongs to metals which play an important role in body metabolism and is not toxic. The Mn, Mg, Co, Cu, Zn, and Ni are involved in growth, development, and reproduction process, but in high concentrations can show toxic effects (see the above section *"Laboratory tests—acute and chronic effects"*). Pb and Cd are not involved in metabolic processes; thus, they are highly toxic at low concentrations and have a great storage capacity in the organisms at long-term exposure. The results on metal concentrations in *U. pictorum* and *A. cygnea* shells, metal detection in sediment samples (2013), average of metals detection in sediment in period of 2009–2013 at Uzlina and BSAFsed values are presented in

Table 8. The metals were determined in both bivalve species *U. pictorum* and *A. cygnea* shells, in September 2013 (**Table 8**).

Metal	Cb* *Unio pictorum*	BSAFsed	Cb* *Anodonta cygnea*	BSAFsed	Csed*	Csed* 2009–2013**
As	<0.05	0.007		0.008	6.60	9.30
Cd	<0.01	–	<0.01	–	–	0.51
Cu	2.57	0.05	4.63	0.09	48.9	47.0
Cr	<0.01	0.0004	0.14	0.005	28.2	29.2
Co	0.13	0.01	<0.01	0.001	9.47	9.84
Fe	153	–	97.8	–	–	20,987
Mn	248	–	157	–	–	614
Ni	0.29	0.008	0.24	0.007	35.6	39.3
Pb	<0.05	0.002	<0.05	0.001	30.2	21.3
Se	<0.09	–	<0.09	–	–	–
Sb	<0.05	–	<0.05	–	–	–
Zn	0.59	0.006	1.27	0.01	91.2	96.9
Mg	42.8	–	79.6	–	–	–

* Average of metal concentrations (for two replicates) expressed in mg/kg d.m.
** Csed 2009–2013—average of the metal concentration detected in sediment from 2009 to 2013 [63, 65].

Table 8. Metal concentration (mg/kg d.m) in the shell of *Unio pictorum* and *Anodonta cygnea* at Uzlina in September 2013.

In this case, the highest metal concentration was recorded for Cu, Ni, and Zn. The Cu and Ni concentrations from sediment exceed the allowed limit values both in September 2013 and as well as during 2009–2013 monitoring period. As shown in **Table 8**, the *A. cygnea* were found to have a greater capacity for metal accumulation (especially for Cu, As, Cr, Zn) than *U. pictorum* shells.

The bioaccumulation level varied depending on species, metals type, and sampling sites. No significant differences were observed between bioaccumulation factors of Cu, Zn, and Ni calculated for *U. pictorum* collected in July and September. The BSAFsed values were subunitary maintained. It was observed a difference considering the sampling points, respectively, at Murighiol the bioaccumulative metals impact (Ni and Zn) was greater compared to Uzlina. This aspect may be explained by the dredging works for the canal enlarging/widening to facilitate navigation, allowing a better water circulation from the branch inside the canal.

This preliminary study for the metal bioaccumulation capacity in the shell mollusks from Danube Delta aquatic system showed that essential metals involved in metabolic processes (such as Fe, Mn, Zn, Cu, and Mg) have a greater storage capacity than those toxic (such as Pb and Cd). The statement was also confirmed in other studies [83–85].

All the biota sediment bioaccumulation factors were subunitary, which indicated a slowly bioaccumulation process occurred in the studied aquatic ecosystems.

3.4. Preliminary risk assessment

Risk characterization is required for all chemicals as an estimation of their exposure and adverse effects on the environmental compartment. Generally, this is based on Predicted Environmental Concentration (PEC) and Predicted No Effect Concentration (PNEC) calculation, in terms of exposure and assessment of effects [86].

In order to estimate the current contamination of Danube surface water and sediment with metals, we use the average of the measured environmental concentrations (MEC) as PEC values, for the period 2009–2013 at Murighiol and Uzlina. The PNEC value calculation was made using an assessment factor (AF) of 1000 applied for acute toxicity values—LC50 (96 h) or 10 applied for chronic toxicity values—MATC for *C. carpio* (our laboratory tests), which expresses the degree of uncertainty in the actual environmental extrapolation [87]. The risk quotients (RQs) between MEC values and acute or chronic PNECs were calculated, and the level of risk was expressed as: insignificant risk (RQs <0.1); low risk (RQs <1); moderate risk (RQs <10), and high risk (RQs >10). The estimated RQs for the most detected metals in Danube water and sediment (Ni, Cd, Cr, Cu, Pb, and Zn) were summarized in **Table 9**.

Metal	MEC (µg/L)*		PNEC (µg/L)		RQs acute		Risk level	RQs chronic		Risk level
	S7	S8	Acute (AF = 1000)	Chronic (AF = 10)	S7	S8		S7	S8	
Surface water										
Ni	12.4	2.60	65.8	10.0	0.18	0.03	L/I	1.24	0.26	L
Cd	0.36	0.37	0.16	0.10	2.25	2.31	M	3.60	3.70	M
Cr	5.38	3.52	120	100	0.04	0.02	I	0.05	0.04	I
Cu	12.9	14.5	2.17	5.00	5.94	6.68	M	2.58	2.90	M
Pb	2.15	2.17	30.1	100	0.07	0.07	I	0.02	0.02	I
As	1.82	1.73	0.40	0.50	4.55	4.32	M	3.64	3.46	M
Zn	9.58	8.15	12.2	60.0	0.78	0.66	L	0.16	0.14	L
Sediment										
Ni	30.8	39.3	65.8	10.0	0.46	0.59	L	3.08	3.93	M
Cd	0.50	0.51	0.16	0.10	3.12	3.18	M	5.00	5.10	M
Cr	31.6	29.2	120	100	0.26	0.24	L	0.32	0.29	L
Cu	35.1	47.0	2.17	5.00	16.2	21.7	H	7.01	9.41	M
Pb	22.3	21.3	30.1	100	0.74	0.70	L	0.22	0.21	L
As	9.61	9.30	0.40	0.50	24.0	23.3	H	19.2	18.6	H
Zn	88.5	96.9	12.23	60.0	7.23	7.92	M	1.48	1.62	M

* Average of concentrations in period of 2009–2013; I—insignificant risk; L—low risk; M—moderate risk; H—high risk.

Table 9. Estimated acute and chronic RQs at Murighiol (S7) and Uzlina (S8) for *Cyprinus carpio*.

The results showed different levels of risk in accordance with detected environmental concentration of metals, the acute and chronic toxicity and the environmental compartment (water or sediment). In water, Cr and Pb showed insignificant risk; Ni and Zn showed a low risk; and Cd, Cu, and As highlight a moderate risk considering both acute and chronic effects on *C. carpio*. Variations of the RQs depending on sampling location are not observed.

As we expected, the risk level increases within the sediment compartment. The sediment contamination revealed low-to-moderate risk, exception for As and Cu. Therefore, Cr and Pb showed low risk; Ni, Cd, Zn and Cu highlighted moderate risk; and As and Cu could express a high risk on fish *C. carpio*. Cu, Zn, and Ni were constantly present in sediment over the set limits of national norms inducing also their accumulation (see the section "*Field tests— bioaccumulation*"). No variation is observed of the RQs depending on sampling location. Using long-term toxicities in PNECs estimation, the RQs increased for Ni, Cd, and Cr and decreased in case of Cu, Pb, As, and Zn, due to the use of a small applied factor (AF = 10) to chronic toxicities.

The results highlighted a pessimistic view concerning the quality of aquatic ecosystem needed to support the carp fish survival. The concern is related to the constantly presence of metal concentrations especially in sediments (the food provider compartment) which could determinate the bioaccumulation. The same statement was made in a Romanian study named "*Ecotoxicology of heavy metals in Danube meadow*" [55].

4. Conclusions

The topic of this chapter was based on the assessment of aquatic systems quality related to persistent metal pollution. The toxic metals are the most frequently detected pollutants in the aquatic environmental, and their effects identification are essential to protect the ecosystems integrity as well as human health. Metal pollution is a global problem; thus, the international regulations with regard to the water quality demand compliance with the quality standards in surface water, groundwater, and biota. The use of organisms (such as fish, crustacean, and mollusks) as bioindicators of metal pollution allowed us to obtain valuable information about the effects on the Romanian common species and to estimate the quality of their environment. The results from laboratory toxicity tests showed the highest concentration values that are not relevant for the detected metal concentrations into surface water, but the metals accidentally released and long-term accumulation could create similar conditions to the results of applied tests. Cd, As, Cu, Zn, Pb, Ni, Zr, and Ti have a very toxic and toxic effects for *C. carpio* and could raise concerns because of its importance for human as a fishery resource. Benthic invertebrates' analysis of the bioaccumulation level varied between species, metals type, and sampling sites. The metal analysis in mollusks shell showed that the metals involved in the metabolic processes (Fe, Mn, Zn, Cu, and Mg) had greater storage capacity than the toxic one (Pb, Cd). In case of *V. viviparus* shell, the selectivity of the metal concentration was represented as follows: Fe > Mn > Zn > Cu > Pb > Co > Cd, while the shell of *A. cygnea* had a greater accumulation capacity for Cu, As, Cr, Zn compared to *Unio* sp. The bioaccumulation factors of metals in benthic organisms were subunitary, which indicated a slowly bioaccumulation process occurred in the studied aquatic ecosystems. This conclusion highlighted a bioaccumulation process that can increase the persistence of metals in the ecosystem, with a long-term potential risk in trophic chain. The preliminary aquatic risk assessment calculated for *C. carpio* for the most detected metals both in water and in sediment (Ni, Cd, Cr, Cu, Pb, As, and Zn) revealed insignificant to moderate risk considering the metals measured environmental concentrations, acute and long-term effects. The results highlighted a pessimistic view concerning the quality of aquatic ecosystem needed to support the carp survival. The concern is related to the constant presence of metal concentrations especially in sediments which is the principal food provider, leading to bioaccumulation processes and trophic chain transfer. Future studies have been initiated to understand the long-term effects of metals in aquatic biota and to complete the aquatic risk assessment considering the abiotic factors.

Abbreviations:

Cd	cadmium
As	arsenium
Cu	copper
Pb	lead
Ni	nickel
Zr	zirconium
Ti	titanium
Fe	iron
Zn	zinc
Mn	manganese
Mg	magnesium
Cd	cadmium
Co	cobalt
Cr	chromium
Mo	molybdenum
Se	selenium
Na	sodium
P	phosphorus

S	sulfur
Hg	mercury
CN	cyanide
LC (EC) 50	lethal concentrations for 50% of tested organisms after 96 or 48 h
MATC	maximum acceptable toxicant concentration in aquatic systems
NOEC	no observed effect concentration
LOEC	low observed effect concentration
GOT	glutamic oxaloacetic transaminase
GPT	glutamic pyruvic transaminase
OECD	Organization for Economic Co-operation and Development
PNEC	predicted no-effect concentration
PEC	predicted exposure concentration
MEC	measured environmental concentration
RQ	risk quotient

Author details

Stefania Gheorghe*, Catalina Stoica, Gabriela Geanina Vasile, Mihai Nita-Lazar, Elena Stanescu and Irina Eugenia Lucaciu

*Address all correspondence to: biologi@incdecoind.ro

National Research and Development Institute for Industrial Ecology – ECOIND, Bucharest, Romania

References

[1] Sandu C, Farkas A, Musa-Iacob R, Ionica D, Parpala L, Zinevici V, Dobre D, Radu M, Presing M, Casper H, Buruiana V, Wegmann K, Stan G, Bloesch J, Triebskorn R, Köhler H-R. Monitoring pollution in River Mures, Romania, part I: the limitation of traditional methods and community response. Large Rivers 2008; 18(1–2): 91–106. doi:10.1127/lr/18/2008/91.

[2] Zhang W, Zhang Y, Zhang L, Lin Q. Bioaccumulation of metals in tissues of seahorses collected from Coastal China. Bull Environ Contam Toxicol 2016; 96(3): 281–288. doi: 20.1007/s00128-16-1728-4

[3] Stankovic S, Kalaba P, Stankovic RA. Biota as toxic metal indicators. Environ Chem Lett 2014; 12: 63–84. doi:10.1007/s10311-013-0430-6

[4] Velma V, Tchounwou PB. Chromium-induced biochemical, genotoxic and histopatho-logic effects in liver and kidney of goldfish, *Carassius auratus*. Mutat Res 2010; 698 (1–2): 43–51. doi:10.1016/j.mrgentox.2010.03.014

[5] Conceiçao Vieira M, Torronteras R, Córdoba F, Canalejo A. Acute toxicity of manganese in goldfish *Carassius auratus* is associated with oxidative stress and organ specific antioxidant responses. Ecotoxicol Environ Saf 2012; 78: 212–217. doi:10.1016/j.ecoenv.2011.11.015

[6] WFW Directive 2000/60/EC of the European Parliament and of the Council of 23 October 2000 establishing a framework for Community action in the field of water policy.

[7] Directive 2008/1/EC of the European Parliament and of the Council of 15 January 2008 concerning integrated pollution prevention and control.

[8] Directive 2008/56/EC of the European Parliament and of the Council of 17 June 2008 establishing a framework for community action in the field of marine environmental policy (Marine Strategy Framework Directive).

[9] Commission Regulation (EC) No 629/2008 of 2 July 2008 amending Regulation (EC) No 1881/2006 setting maximum levels for certain contaminants in foodstuffs (Text with EEA relevance).

[10] El-Moselhy KhM, Othman AI, Abd El-Azem H, El-Metwally MEA. Bioaccumulation of heavy metals in some tissues of fish in the Red Sea, Egypt. Egypt J Basic Appl Sci 2014; 1(2): 97–105. doi:10.1016/j.ejbas.2014.06.001

[11] Moloukhia H, Sleem S. Bioaccumulation, fate and toxicity of two heavy metals common in industrial wastes in two aquatic mollusks. J Am Sci 2011; 7(8): 459–464. ISSN 1545-1003.

[12] Stankovic S, Jovic M. Health risks of heavy metals in the Mediterranean mussels as seafood. Environ Chem Lett 2012; 10(2): 119–130. doi:10.1007/s10311-011-0343-1

[13] Stankovic S, Jovic M. Native and invasive mussels. In: Nowak J, Kozlowski M (eds). Mussels: ecology, life habits and control. Chapter 1. NOVA Publisher: NY; 2013. p. 1–45. ISBN: 978-1-62618-084-0.

[14] Stankovic S, Stankovic RA. Bioindicators of toxic metals. In: Lichtfouse E et al (eds). Green materials for energy, products and depollution, environmental chemistry for a sustainable world, 3rd ed. Chapter 5. Springer: Berlin; 2013, p.151–228. doi: 10.1007/978-94-007-6836-9_5

[15] Hauser-Davis RA, Calixto de Campos R, Ziolli RL. Fish metalloproteins as biomarkers of environmental contamination. Rev Environ Contam Toxicol 2012; 218: 101–123. doi: 10.1007/978-1-4614-3137-4_2

[16] Frémion F, Bordas F, Mourier B, Lenain JF, Kestens T, Courtin-Nomade A. Influence of dams on sediment continuity: a study case of a natural metallic contamination. Sci Total Environ 2016; 547: 282–294. doi:10.1016/j.scitotenv.2016.01.023.

[17] Hasan MR, Khan MZH, Khan M, Aktar S, Rahman M, Hossain F, Hasan ASMM. Heavy metals distribution and contamination in surface water of the Bay of Bengal coast. Environ Sci 2016; 2(1): 1–12. doi:10.1080/23311843.2016.1140001

[18] Varol M, Şen B. Assessment of nutrient and heavy metal contamination in surface water and sediments of the upper Tigris River, Turkey. Catena 2012; 92: 1–10. doi:10.1016/j.catena.2011.11.011

[19] Equeenuddin SM, Tripathy S, Sahoo P, Panigrahi M. Metal behavior in sediment associated with acid mine drainage stream: role of pH. J Geochem Explor 2013; 124: 230–237. doi:10.1016/j.gexplo.2012.10.010

[20] Vasile G, Cruceru L, Petre J, Iancu V. Complex analytical investigations regarding the bio-availability of heavy metals from sediments. Rev Chim (Bucharest) 2005; 56(8): 790–794.

[21] Griscom SB, Fisher NS. Bioavailability of sediment-bound metals to marine bivalve mollusks: an overview. Estuaries 2004; 27(5): 826–838. doi:10.1007/BF02912044

[22] Roosa S, Prygiel E, Lesven L, Wattiez R, Gillan D, Ferrari BJ, Criquet J, Billon G. On the bioavailability of trace metals in surface sediments: a combined geochemical and biological approach. Environ Sci Pollution Res 2016; 23(11): 10679–10692. doi:10.1007/s11356-016-6198-z

[23] Fu J, Zhao C, Luo Y, Liu C, Kyzas GZ, Luo Y, Zhao D, An S, Zhu H. Heavy metals in surface sediments of the Jialu River, China: their relations to environmental factors. J Hazard Mater 2014; 270: 102–109. doi:10.1016/j.jhazmat.2014.01.044

[24] Bonnail E, Sarmiento AM, DelValls TA, Nieto JM, Riba I. Assessment of metal contamination, bioavailability, toxicity and bioaccumulation in extreme metallic environments (Iberian Pyrite Belt) using Corbicula fluminea. Sci Total Environ 2016; 544: 1031–1044. doi:10.1016/j.scitotenv.2015.11.131

[25] Bryan GW, Langston WJ, Hummerstone LG and Burt GR. A guide to the assessment of heavy metal contamination in estuaries using biological indicators. Mar Biolo Assoc UK 1985; 4: 91–110.

[26] Roesijadi G, Robinson WE. Metal regulation in aquatic animals: mechanisms of uptake, accumulation, and release. In: Malins DC and Ostrander GK (eds). Aquatic toxicology: molecular, biochemical, and cellular perspectives. Lewis Publishers: Boca Raton; 1994. p. 387–420. ISBN-13: 978-0873715454. ISBN-10: 0873715454.

[27] Wojtkowska M, Bogacki J, Witeska A. Assessment of the hazard posed by metal forms in water and sediments. Sci Total Environ 2016; 551–552: 387–392. doi:10.1016/j.scitotenv.2016.01.073

[28] Wang W.X, Fisher NS. Modeling metal bioavailability in marine mussels. Rev Environ Contam Toxicol 1997; 151: 39–65. doi:10.1007/978-1-4612-1958-3_2

[29] Chapman PM, Wang F, Janssen C, Persoone G, Allen HE. Ecotoxicology of metals in aquatic sediments: binding and release, bioavailability, risk assessment, and remediation. Can J Fish Aquat Sci 1998 (on-line 2011); 55(10): 2221–2243. doi:10.1139/f98-145

[30] Bryan GW. Some aspects of heavy metals tolerance in aquatic organisms. In: Lockwood APM (ed). Effects of pollutants on aquatic organisms. Cambridge University Press: Cambridge; 1976. p. 7–35.

[31] Eggleton J, Thomas KV. A review of factors affecting the release and bioavailability of contaminants during sediment disturbance events. Environ Int 2004; 30(7): 973–980. doi:10.1016/j.envint.2004.03.001

[32] Adams WJ, Chapman PM. Assessing the hazard of metals and inorganic metal substances in aquatic and terrestrial systems. New York, USA; CRC Press; 2007. ISBN 9781420044416.

[33] Rosado D, Usero J, Morillo J. Assessment of heavy metals bioavailability and toxicity toward *Vibrio fischeri* in sediment of the Huelva estuary. Chemosphere 2016; 153: 10–17. doi:10.1016/j.chemosphere.2016.03.040

[34] Atici T, Obali O, Altindag A, Ahiska S, Aydin D. The accumulation of heavy metals (Cd, Pb, Hg, Cr) and their state in phytoplanktonic algae and zooplanktonic organisms in Beysehir Lake and Mogan Lake, Turkey. Afr J Biotechnol 2010; 9(4): 475–487. ISSN: 1684-5315.

[35] Bere T, Chia MA, Tundisi J.G. Effects of Cr III and Pb on the bioaccumulation and toxicity of Cd in tropical periphyton communities: implications of pulsed metal exposures. Environ Pollut 2012; 163: 184–191. doi:10.1016/j.envpol.2011.12.028

[36] Brraich OS, Kaur M. Ultrastructural changes in the gills of a cyprinid fish, *Labeo rohita* (Hamilton, 1822) through scanning electron microscopy after exposure to Lead Nitrate (Teleostei: *Cyprinidae*). Iran J Ichthyol 2015; 2(4): 270–279. P-ISSN: 2383-1561; E-ISSN: 2383-0964.

[37] Sevcikova M, Modra H, Blahov J, Dobsikova R, Plhalova L, Zitka O, Hynek D, Kizek R, Skoric M, Svobodova Z. Biochemical, haematological and oxidative stress responses of common carp (*Cyprinus carpio* L.) after sub-chronic exposure to copper. Veterinarni Medicina 2016; 61(1): 35–50.

[38] Mishra AK, Mohanty B. Acute toxicity impacts of hexavalent chromium on behavior and histopathology of gill, kidney and liver of the freshwater fish, *Channa punctatus* (Bloch). Environ Toxicol Pharmacol 2008; 26(2): 136–141. doi:10.1016/j.etap.2008.02.010

[39] Eroglu A, Dogan Z, Kanak EG, Atli G, Canli M. Effects of heavy metals (Cd, Cu, Cr, Pb, Zn) on fish glutathione metabolism. Environ Sci Pollut Res 2015; 22(5): 3229–3237. doi: 10.1007/s11356-014-2972-y

[40] Jiang H, Kong X, Wang S, Guo H, Effect of copper on growth, digestive and antioxidant enzyme activities of Juvenile Qihe Crucian Carp, *Carassius carassius*, during exposure and recovery. Bull Environ Contam Toxicol 2016; 96(3): 333–340. doi:10.1007/s00128-016-1738-2

[41] El Basuini MF, El-Hais AM, Dawood MAO, El-Sayed Abou-Zeid A, EL-Damrawy SZ, EL-Sayed Khalafalla MM, Koshio S, Ishikawa M, Dossou S. Effect of different levels of dietary copper nanoparticles and copper sulfate on growth performance, blood biochemical profiles, antioxidant status and immune response of red sea bream (*Pagrus major*). Aquaculture 2016; 455: 32–40. doi:10.1016/j.aquaculture. 2016.01.007

[42] Köhler H-R, Sandu C, Scheil V, Nagy-Petrica EM, Segner H, Telcean I, Stan G, Triebskorn R. Monitoring pollution in river Mureş, Romania, Part III: biochemical effect markers in fish and integrative reflection. Environ Monit Assess 2007; 127(1): 47–54. doi:10.1007/s10661-006-9257-y

[43] Triebskorn R, Telcean I, Casper H, Farkas A, Sandu C, Stan G, Colărescu O, Dori T, Köhler H-R. Monitoring pollution in River Mureş, Romania, part II: metal accumulation and histopathology in fish. Environ Monit Assess 2008; 141(1): 177–188. doi:10.1007/s10661-007-9886-9

[44] Zhou Q, Zhang J, Fu J, Shi J, Jiang G, Biomonitoring: an appealing tool for assessment of metal pollution in the aquatic ecosystem. Anal Chim Acta 2007; 14(606): 135–150. doi:10.1016/j.aca.2007.11.018

[45] PAN Pesticides Database, http://www.pesticideinfo.org.

[46] Pickering QH, Gast MH. Acute and chronic toxicity of cadmium to the fathead minnow (*Pimephales promelas*). J Fish Res Board Canada 1972 (on-line 2011); 29(8): 1099–1106. doi:10.1139/f72-164

[47] Rehwoldt R, Menapace LW, Nerrie B, Allessandrello D. The effect of increased temperature upon the acute toxicity of some heavy metal ions. Bull Environ Contam Toxicol 1972; 8(2): 91–96.

[48] Schäfer S, Buchmeier G, Claus E, Duester L, Heininger P, Körner A, Mayer P, Paschke A, Rauert C, Reifferscheid G, Rüdel H, Schlechtriem C, Schröter-Kermani C, Schudoma D, Smedes F, Steffen D, and Vietoris F. Bioaccumulation in aquatic systems: methodological approaches, monitoring and assessment. Environ Sci Eur 2015; 27: 5. doi:10.1186/s12302-014-0036-z

[49] Kominkova D, Nabelkova J. Effect of urban drainage on bioavailability of heavy metals in recipient. Water Sci Technol 2007; 56(9): 43–50. doi:10.2166/wst.2007.736

[50] Schneider L, Belgerb L, Burgerc J, Vogta RC. Mercury bioaccumulation in four tissues of *Podocnemis erythrocephala* (Podocnemididae: Testudines) as a function of water parameters. Sci Total Environ 2009; 407: 1048–1054. doi:10.1016/j.scitotenv. 2008.09.049

[51] Salazar MJ, Rodriguez J H, Nieto GL, Pignata ML. Effects of heavy metal concentrations (Cd, Zn and Pb) in agricultural soils near different emission sources on quality, accumulation and food safety in soybean [*Glycine max* (L.) Merrill]. J Hazard Mater 2012; 233–234: 244–225. doi:10.1016/j.jhazmat.2012.07.026

[52] Zhu X, Chang Y, Chen Y., Toxicity and bioaccumulation of TiO_2 nanoparticle aggregates in *Daphnia magna*. Chemosphere 2010; 78: 209–215. doi:10.1016/j.chemosphere.2009

[53] Gobas FAPC. Assessing bioaccumulation factors of persistent organic pollutants in aquatic food-chains. In: Stuart H (ed) Persistent organic pollutants. Springer: US; 2001. p. 145–165. doi:10.1007/978-1-4615-1571-5_6

[54] Vădineamu A. "Considerations on the significance of holistic approach of the heavy metals and radioactive pollution problems. Nature Protection". Ocrotirea Naturii 1990; 1(1): 51–54.

[55] Iordache V. Ecotoxicology of heavy metals in Danube meadow. Bucharest, Romania; Ars Docenti; 2009. ISBN 978-973-558-233-3.

[56] Jitar O, Teodosiu C, Oros A, Plavan G, Nicoara M. Bioaccumulation of heavy metals in marine organisms from the Romanian sector of the Black Sea. N Biotechnol 2015; 25; 32(3): 369–378. doi:10.1016/j.nbt.2014.11.004

[57] Vinodhini R, Narayanan R. Bioaccumulation of heavy metals in organs of fresh water fish *Cyprinus carpio* (Common carp). Int J Environ Sci Tech 2008; 5(2): 179–182. doi: 10.1007/BF03326011

[58] Baby J, Raj SJ, Biby ET, Sankarganesh P, Jeevitha MV, Ajisha Su and RAJAN SS. Toxic effect of heavy metals on aquatic environment. Int J Biol Chem Sci 2010; 4(4): 939–952. ISSN 1991-8631.

[59] Saha N, Zaman MR. Evaluation of possible health risks of heavy metals by consumption of foodstuffs available in the central market of Rajshahi City, Bangladesh. Environ Monit Assess 2013; 185(5): 3867–3878. doi:10.1007/s10661-012-2835-2

[60] Ashish T, Amitabh CD. Assessment of heavy metals bioaccumulation in alien fish species, *Cyprinus carpio* from the Gomti River, India. Eur J Exp Biol 2014; 4(6): 112–117. ISSN: 2248-9215.

[61] David IG, Matache ML, Tudorache A, Chisamera Gabriel, Rozylowicz L, Radu GL. Food chain biomagnification of heavy metals in samples from the lower Prut floodplain natural park. Environ Eng Manag J 2012; 11(1): 69–73.

[62] Rainbow PS, Poirier L, Smith BD, Brix KV, Luoma SN. Trophic transfer of trace metals from the polychaete worm Nereis diversicolor to the polychaete *N. virens* and the decapod crustacean *Palaemonetes varians*. Mar Ecol Prog Ser 2006; 321: 167–181.

[63] Stanescu E, Stoica C, Vasile G, Petre J, Gheorghe S, Paun I, Lucaciu I, Nicolau M, Vosniakos F, Vosniakos K, Golumbeanu M. Structural changes of biological compartments in Danube Delta systems due to persistent organic pollutants and toxic metals. In: L.I. Simeonov et al. (eds). Environmental security assessment and management of obsolete pesticides in Southeast Europe, NATO Science for Peace and Security Series C: Environment Security. Springer Science+ Business Media: Dordrecht 2013, p. 229–248. doi:10.1007/978-94-007-6461-3_21

[64] Stoica C, Stanescu E, Lucaciu I, Gheorghe S, Nicolau M. Influence of global change on biological assemblages in the Danube Delta. J Environ Prot Ecol 2013; 14(2): 468–479.

[65] Stoica C, Gheorghe S, Paun I, Stanescu E, Dinu C, Petre J, Lucaciu I. Long term biological changes along Danube Delta system after industrialization period. Revista Rom Aqua 2014; 1: 14–20.

[66] Stoica C, Gheorghe S, Lucaciu I, Stanescu E, Paun I, Niculescu D. The impact of chemical compounds on benthic invertebrates from Danube–Danube Delta systems. Soil Sediment Contam Int J 2014; 23(7): 763–778. doi:10.1080/15320383.2014.870529

[67] Cordos E, Rautiu R, Roman C, Ponta M, Frentiu T, Sarkany A, Fodorpataki L, Macalik K, McCormick C, Weiss D. Characterization of the rivers system in the mining and industrial area of Baia Mare, Romania. Eur J Mineral Process Environ Prot 2003; 3(3) 1303–0868: 324–335.

[68] Dumitrel GA, Glevitzky M, Popa M, Vica ML. Studies regarding the heavy metals pollution of streams and rivers in Rosia Montana area, Romania. J Environ Prot Ecol 2015; 16(3): 850–860.

[69] Iordache M, Popescu LR, Pascu LF, Iordache I. Environmental risk assessment in sediments from Jiu River, Romania. Rev Chim (Bucharest) 2015; 66(8): 1247–1252.

[70] Alam MK, Maughan OE. The effect of malathion, diazinon, and various concentrations of zinc, copper, nickel, lead, iron, and mercury on fish. Biol Trace Elem Res 1992; 34(3): 225–236.

[71] EaSI-Pro® View 14.0 Data base for dangerous chemicals, 2005 Haskoning B.V. Available from: http://www.ekotox.eu/component/content/article/118-EASI-pro-view.

[72] Rand MG. Fundamentals of aquatic toxicology: effects, environmental fate and risk assessment, 2nd ed. North Palm Beach, Florida. CRC Press 1995; p. 943. ISBN1-56032-090-7.

[73] Besser JM, Leib KJ. Toxicity of metals in water and sediment to aquatic biota. In: Stanley E. Church, Paul von Guerard, and Susan E. Finger (eds). Integrated investigations of

environmental effects of historical mining in the Animas River watershed. San Juan County. Colorado; 2007. p. 839–849.

[74] Stoica C, Gheorghe S, Petre J, Lucaciu I, Nita-Lazar M, Tools for assessing Danube Delta systems with macro invertebrates. Environ Eng Manag J 2014; 13(9): 2243–2252.

[75] Kesavan K, Murugan A, Venkatesan V, Vijay Kumar B.S. Heavy metal accumulation in molluscs and sediment from uppanar estuary, Southeast coast of India. Thalassas. Int J Mar Sci 2013; 29(2): 15–21.

[76] Findlater G, Shelton A, Rolin T, Andrews J. Sodium and strontium in mollusc shells: preservation, palaeosalinity and palaeotemperature of the Middle Pleistocene of eastern England. Proc Geololists Assoc 2014; 3125(1): 14–19. doi:10.1016/j.pgeola.2013.10.005

[77] Kadar E, Costa V. First report on the micro-essential metal concentrations in bivalve shells from deep-sea hydrothermal vents. J Sea Res 2006; 56(1): 37–44. doi:10.1016/j.seares.2006.01.001

[78] Eisler R. Molluscs. In: Compendium of trace metals and marine biota. 1st ed. Elsevier: Amsterdam; 2010, p. 143–397 ISBN: 978-0-444-53439-2.

[79] Contia ME, Finoia MG. Metals in molluscs and algae: a north–south Tyrrhenian Sea baseline. J Hazard Mater 2010; 181(1–3): 388–392. doi:10.1016/j.jhazmat.2010.05.022

[80] Beek B. Bioaccumulation: new aspects and developments. In: Otto Hutzinger, editor. Handbook of environmental chemistry. 2nd ed. Reactions and processes, Part J. Springer-Verlag: New York; 2000. p. 284. doi:10.1007/10503050

[81] Zver'kova YS. Use of freshwater Mollusk Shells for monitoring heavy metal pollution of the dnieper ecosystem on the territory of Smolensk Oblast. Russ J Ecol 2009; 40(6): 443–447. doi:10.1134/S1067413609060113

[82] Gupta S.K, Singh J. Evaluation of mollusc as sensitive indicator of heavy metal pollution in aquatic systems: a review. IIOAB J Special Issue Environ Manag Sustain Dev 2011; 2(1):49-57.

[83] Shaari H, Raven B, Sultan K, Mohammad Y, Yunus K. Status of heavy metals concentrations in oysters (Crassostrea sp.) from Setiu Wetlands, Terengganu, Malaysia. Sains Malaysiana 2016; 45(3): 417–424.

[84] Oros A, Gomoiu M-T. Comparative data on the accumulation of five heavy metals cadmium, chromium, copper, nickel, lead) in some marine species (mollusks, fish) from the Romanian of the Black Sea 2010 (http://www.rmri.ro/Home/Downloads/Publications.RecherchesMarines/2010/paper03.pdf).

[85] Yusoff N.A.M, Long S.M. Comparative bioaccumulation of heavy metals (Fe, Zn, Cu, Cd, Cr, and Pb) in different edible mollusk collected from the estuary area of Sarawak

River 2011; UMTAS Empowering Science, Technology and Innovation Towards a Better Tomorrow. p. 806–811.

[86] TDG (2003). Technical Guidance Document on Risk Assessment, Commission Directive 93/67/EEC on Risk Assessment for new notified substances, European Commission.

[87] Gheorghe S, Lucaciu I, Paun I, Stoica C, Stanescu E. Environmental exposure and effects of some micropollutants found in Romanian surface waters. J Environ Prot Ecol 2014; 5(3): 2014.

8

Ecosystem Approach to Managing Water Quality

Oghenekaro Nelson Odume

Additional information is available at the end of the chapter

Abstract

This chapter argues for the ecosystem approach to managing water quality, which advocates the management of water, land and the associated living resources at the catchment scale as complex social-ecological systems and proactively defend and protect the ecological health of the ecosystem for the continuing supply of ecosystem services for the benefit of society. It argues for a shift from the engineering-driven command and control approach to water resource management. Environmental water quality (EWQ) is discussed as a holistic and integrated tripod ecosystem approach to managing water quality. Water physico-chemistry, biomonitoring and aquatic ecotoxicology are discussed as and their application and limitation with respect to water quality management, particularly in South Africa, is critically evaluated. The chapter concludes with a case study illustrating the application of biomonitoring for the assessment of ecosystem health in the Swartkops River, Eastern Cape, South Africa. The macroinvertebrates-based South African Scoring System version 5 was applied at three impacted sites and one control site. Two of the three impacted sites downstream of an effluent discharge point had very poor health conditions. The urgent need for ecological restoration was recommended.

Keywords: aquatic ecosystems, biomonitoring, ecotoxicology, macroinvertebrates, pollution, water chemistry

1. Introduction

The sustainability of freshwater ecosystems is being threatened globally [1]. A growing human population, coupled with changing demography, increasing socio-economic development as well as urbanisation and industrialisation of freshwater ecosystems catchments are the major drivers of change, resulting in deteriorating freshwater quality and depleting quantity. Climate change and other human-induced influences will, in the foreseeable future, exacerbate the conditions of the already stressed freshwater ecosystems [2]. Globally, there is a growing recognition that the typical hard-engineering informed 'command-and-control' approach to managing freshwater ecosystems, particularly water quality, is no longer sustainable [3, 4]. The hard engineering command and control approach (CCP) arises out of the insatiable quest for humans to tame, control and command everything in the environment, including nature [4]. Its primacy is the development of water resources for the socio-economic benefits of human with little or no attention to the ecosystems that provide the resource base. It is, however, becoming increasingly clear that an alternative approach that takes account of both ecosystem sustainability and socio-economic development is needed for managing water resources, including water quality.

The ecosystem approach is a holistic and integrated management strategy with an apprecia-tion of the ecosystem as the source of water as well as a water user with specific requirements in terms of water quality, quantity, in-stream ecological and riparian conditions as well as the overall health and functionality of the ecosystem [5]. It advocates the management of water, land and the associated living resources at the catchment scale as complex social-ecological systems [6]. It proactively defends and protects the ecological health of the ecosystem. It is becoming the preferred approach for managing water quality, for example, in Europe [7], Australia [8] and South Africa [9]. For example, the European Union Water Framework Direc-tive (WFD, 2000/60/EC) explicitly recognises and consciously advocates the ecosystem approach to managing the surface water quality of water bodies within the EU member states. It mandates all EU members to maintain surface water quality in 'good status' and to restore degraded systems to 'good conditions'.

2. The ecosystem approach and water quality in South Africa

South Africa's ground-breaking water law provides for an ecosystem approach to managing water resources (National Water Act No. 36 of 1998). The strategies for achieving the ecosys-tem-oriented objectives of the Act are designed in the National Water Resource Strategy 2 (NWRS2) [5]. The NWRS2 provides for two complementary approaches, the Resource Directed Measures (RDM) and the Source Directed Controls (SDC).

The RDM are directed at protecting and using the water resources sustainably, in terms of water quality, ecological and riparian habitat conditions [5]. The RDM are composed of the national water resource-classification system, the ecological reserve, and the Resource Quality Objectives (RQOs). In South Africa, water resources are classified into three management classes: Class I (a resource with no noticeable or with minimal human impacts); Class II (a resource slightly or moderately impacted by human activities with little deviation from natural conditions); and Class III (a resource with significant impacts resulting in serious deviation from natural conditions) [5, 10]. Water resources in Classes I and II are given high management priority to keep them in good condition; while depending on the scenarios, efforts are made to restore the conditions of those in management Class III. The ecological reserve provides the legal basis for assessing and protecting the quality, quantity and reliability of water needed for the functioning and maintaining the aquatic ecosystem [9]. The RQO provides measurable quantitative and qualitative descriptions/objectives for the physical, bio-logical and chemical attributes that should be protected. The RQOs thus capture the management class and the ecological requirements, giving directions on how a water resource should be managed to protect key ecosystem attributes and functionalities [11]. The determination of the ecological reserve involves derivation of the present ecological state (PES) of the water resource. The PES is determined by integrating biological, physical and chemical information, including fish, macroinvertebrates, geomorphology, vegetation, riparian condition as well as hydrological and physico-chemical variables.

The NWRS2 also provides for measures to control the use of water resources to protect the water quality and ecological conditions needed to ensure the functionality of the aquatic ecosystems. Human activities impacting water quality in terms of abstraction and discharges are regulated through the SDC, which are used in combination with the RDM. The SDC define, and then impose limits, and restrict the use of water resources to achieve the desired levels of protection. Licensing, registration, authorisation and special permit are the tools used to achieve the control of water use impact on water quality. Guidelines and limits, discharges of effluent as well as water abstractions are used to impose limit on water use activities. The combined process of the RDM and the SDC involves integrating biophysical information from multiple components of the ecosystems, and in terms of water quality, environmental water quality (EWQ) provides a sound ecosystem-based methodology for managing aquatic ecosystems in South Africa.

3. Environmental water quality (EWQ)

Environmental water quality is an integrated approach that links the chemical, physical and radiological characteristics of a water resource to the responses of the in-stream assemblage structure, function and processes [12, 13]. The EWQ combines water physico-chemistry, biomonitoring and ecotoxicology. The conventional approach to managing water quality is physico-chemistry, which involves measuring and analysing physical and chemical variables to indicate water quality without taking into account their effects on biological organisms. Biomonitoring is the systematic deployment of resident biota to provide information on aquatic ecosystem health with limited capacity for a cause-effect relationship, while ecotoxicology is the experimental evaluation of the effects of specific toxicants on aquatic biota, adding the potential for causal linkages.

3.1. Water physico-chemistry

Human activities such as agriculture, domestic and industrial wastewater discharges, environmental engineering, and natural factors including geology and soils, hydrology, seasonal patterns, geomorphology, climate and weather, influence the physico-chemical conditions of the aquatic ecosystems. The physico-chemical variable analysis is the traditional approach to controlling pollution and managing water quality. It helps water-resource managers to measure and analyse the concentrations of pollutants, determine their fate and transport, as well as their persistence in the aquatic environment. In South Africa, for example, the National Physico-Chemical Monitoring Programme (NCPM) uses analyses of physico-chemical variables to provide the water quality status of rivers and streams [14].

The physico-chemical approach forms an important component of the EWQ in terms of managing water quality. However, its drawbacks include (i) high analytical costs of monitoring physico-chemical variables, (ii) inexhaustible numbers of both dissolved and suspended chemicals and pollutants, making the choice of variables for analysis difficult and also making it impossible to measure all variables, (iii) lack of spatial and temporal representativeness of water quality conditions, as results are only reflective of the time and place of sampling and (iv) provision of very little or no insights into ecological response of aquatic biota and overall biophysical health of the system. Given that conserving biodiversity and protecting the ecosystem health are critical objectives of the ecosystem approach, the physico-chemical analysis alone is inadequate. The second pillar of the EWQ, biological monitoring also known as biomonitoring, provides the opportunity for detecting ecological impairments and measuring both taxonomic and functional diversity, which are important components of the aquatic ecosystem.

3.2. Biomonitoring

Biomonitoring integrates multiple effects of stressors including chemical (e.g. salinisation), physical (e.g. sedimentation) and biological (e.g. parasitism) to evaluate aquatic ecosystem health [15]. It relies on the sound ecological understanding that in-stream biota, for example, plants, algae, animals and microorganisms integrate the conditions of their environment and are therefore able to provide an indication of the health of the ecosystem in which they live [16]. Biomonitoring can be applied at multiple biological organisations including sub-organismal (e.g. gene mutation and cell alteration), individual species composition, population, community and ecosystem levels. In South Africa, for example, the science of biomonitoring is well developed compared to the rest of sub-Saharan African countries. The design of the National Aquatic Ecosystem Health Monitoring Programme (NAEHMP) is met to generate information needed regarding the ecological conditions of aquatic ecosystems in South Africa [17]. The NAEHMP utilises the responses of in-stream biota and system drivers to characterise the impacts of disturbances in aquatic ecosystems and to determine present ecological states of the systems. The NAEHMP uses fish, macroinvertebrate and riparian vegetation as its primary biological indicators, while abiotic indicators such as habitat, geomorphology, hydrology and water chemistry form the framework for the interpretation of the biotic results. In terms of the NAEHMP, assessment models such as the fish response assessment index (FRAI), vegetation response assessment index (VEGRAI) and macroinvertebrate response assessment index (MIRAI) have been developed for assessing the ecological states of riverine ecosystems [18–20].

At the core of biomonitoring is the search for and identification of suitable biological indicators (i.e. bioindicators), whose presence or absence, abundance and diversity, and behaviour reflect environmental conditions. Over the years, many studies have used bioindicators such as fish, diatoms, algae and macroinvertebrates to assess ecological water quality [21]. However, among the bioindicators, macroinvertebrates are arguably the most widely used groups [22]. Their wide application in biomonitoring can be attributed to their ubiquitous occurrence, abundance and diversity in the aquatic ecosystems. In addition, they can be easily collected and identified to the family level, though species-level identification requires more time and for some taxa may not be possible especially in the Afrotropical region. They have a huge species richness that offers a wide spectrum of environmental responses and they are relatively sedentary, representing local conditions. They provide an indication of environmental conditions over varying times and are differentially sensitive to a variety of pollutants and, consequently, capable of a graded response to stress. They also serve as a critical pathway for transporting and utilising energy and matter in the aquatic ecosystem.

Freshwater macroinvertebrates spend at least part of their lifecycles in the aquatic environment and are large enough to be seen unaided [23]. Depending on the goal of the biomonitoring, they can be monitored for changes in population, community, growth rate and cohorts. They can also be monitored for bioaccumulation of pollutants, as well as for morphological and biochemical changes in cells, tissues, organs and systems. Macroinvertebrates-based biomonitoring approaches include single biotic indices such as the Biological Monitoring Working Party (BMWP) and the South African Scoring System version 5 (SASS5) [24, 25]; multimetric indices, for example, the Index of Biotic Integrity 12 (IBI 12) and the Serra dos Órgãos Multimetric Index (SOMI) [26]; multivariate predictive techniques, for example, the Australian River Assessment System (AUSRIVAS) and the United Kingdom's River Invertebrate Prediction and Classification System (RIVPACS, UK) [27] and finally the traits-based techniques.

A multivariate predictive technique evaluates aquatic ecosystem condition by comparing biota at a site to those expected to occur in the absence of human disturbances [16]. A predictive model is constructed using reference sites' biotic communities and correlating the community to natural environmental variables using multivariate statistics to predict expected communities at the impacted sites. A multimetric approach on the other hand combines metrics representing several aspects of macroinvertebrate attributes (e.g. structure, function and processes) to indicate river health. Bonada et al. [16] assessed the utilities, strengths and weaknesses of both approaches using a set of 12 criteria in 3 categories: rationale, implementation and performance. Out of the 12 criteria evaluated, the multivariate approach satisfies 9, while the multimetric fulfils 10.

3.3. Aquatic ecotoxicology

Protecting water resources requires a thorough understanding of the mechanisms by which pollutant(s) or toxicant(s) influence the aquatic ecosystems. This often involves experimental manipulation to establish an evidence-based cause-effect relationship between the toxicant and the observed effects on the organism. Aquatic ecotoxicology is the third pillar of the EWQ, and it provides data needed to explore a cause-effect relationship between stressors and biota [28]. The traditional approach to aquatic ecotoxicology is the single-species tests in the laboratory. Depending on the duration of the exposure and the endpoints measured, these tests are termed acute or chronic. Acute toxicity tests are short term, usually lasting between 48 and 96 h, measuring mortality as an endpoint [29]. Chronic toxicity tests last longer, and in addition to long-term mortality, sub-lethal effects on organismal attributes such as growth, reproduction, behaviour, enzymatic activities and histology are also measured. Many of these single-species acute and chronic toxicity tests have been standardised and are widely use in the ecosystem-based approach to managing water quality [30]. The strengths of the laboratory single-species tests include (i) precision: they are conducted in a highly regulated environment, where external influences are isolated, so that there is a high level of precision with regard to the toxicant effects on the organism; (ii) repeatability: single-species, laboratory-based experiments are easily reproducible and repeatable, provided that sets, guidelines and protocols are followed; (iii) high level of acceptance in the regulatory circle: these tests still form the cornerstone of risk assessments of harmful chemicals in the environment; (iv) simplicity: these test are usually very simple to undertake, hence their appeal in regulatory circles.

Although the single-species laboratory-based tests are widely used in managing aquatic ecosystems, they are unable to provide direct community or ecosystem-level effects. They rely heavily on laboratory to field extrapolations by applying safety assessment factors or the species sensitivity distribution (SSD) approach [31]. Reducing uncertainties requires using more ecologically relevant and realistic assessments that employ multi-species in experimental settings that are closer to the natural field conditions. The multi-species model-stream ecosystem approach occupies an intermediate space between field biomonitoring studies and the traditional single-species laboratory-based approach. If reasonably controlled, manipulated and replicated, they can simulate community and even ecosystem effects [32]. While the single-species approach offers high degree of precision, repeatability and simplicity, model-stream ecosystems represent a compromise between these factors, and their high environmental realisms [32].

Model-stream ecosystems are termed mesocosms or microcosms depending on their sizes and locations [33–37]. For example, Odum [33] defined mesocosms as outdoor experimental streams bounded and partially closed, which closely simulate the natural conditions. Buikema and Voshell [34] use volume as a factor for differentiating between microcosms and mesocosms, referring to microcosms as experimental streams (usually indoor) with a volume equal or less than 10 m^3 and mesocosms as those (usually outdoor) having a volume greater than 10 m^3. Hill et al. [35] defined mesocosms as experimental streams that are more than 15 m long and microcosms as those that are shorter. However, Belanger [36] review revealed that increased physical sizes of experimental streams did not correspond to increased biological complexity. Since the goal of a multi-species model-stream ecosystem is to achieve an adequate ecological realism irrespective of size, the terms 'microcosm' and 'mesocosm' are actually inappropriate. Instead, the appropriate terminologies should be a 'model-stream ecosystem', 'experimental streams or artificial streams'.

Model-stream ecosystems have some advantages over conventional single-species toxicity tests. They enable the simulation of natural conditions, offering a high degree of environmental realism and enabling complex biophysical interactions. They enable the researcher to evaluate direct effects of pollutants at higher biological organisation such as population, community and even ecosystem levels [37]. Moreover, they enable the study of biotic interaction and community dynamics and measurement of indirect ecosystem effects. Their shortcomings are that they are not easily reproducible, have low precision and are not simple to undertake.

4. Case study of the application of biomonitoring for the assessment of the ecosystem health in the Swartkops River

This case study illustrates the application of biomonitoring in the Swartkops River using the South African Scoring System version 5.

4.1. The South African Scoring System version 5

The South African Scoring System version 5 (SASS5) is a rapid bioassessment index based on the presence or absence of selected families of aquatic macroinvertebrates and their perceived sensitivity or tolerance to deteriorating water quality [24]. In SASS5, macroinvertebrate families are awarded scores in the range of 1–15 in increasing order of sensitivity to deteriorating water quality. Families considered sensitive are awarded high scores and those considered tolerant low scores. The results are expressed both as an index score, that is, SASS5 score, and as an average score per recorded taxon (ASPT) value. The SASS5 score is calculated by summing the scores of all recorded families, while the ASPT value is obtained by dividing the total SASS5 score by the number of families recorded. In addition to being a useful water quality assessment index, SASS5 is used to assess emerging water quality problems, development impacts, ecological state and spatio-temporal trends of biological assemblages.

4.2. The study area

The Swartkops River originates in the foothills of the Groot Winterhoek Mountains and then meanders through the towns of Uitenhage, Despatch and Perseverance before discharging into the Indian Ocean at Algoa Bay, near the city of Port Elizabeth (**Figure 1**). Climate in the catchment is warm and temperate, and rainfalls vary between the upper and lower regions. The upper region usually receives higher rainfall than the lower region. The catchment geology is mainly of marine, estuarine and fluvial origin. Soils in the upper catchment are not deep and are unsuitable for agriculture. Those in the low-lying floodplain region are deep and well suited for agriculture. The dominant vegetation in the catchment is bushveld and succulent thicket.

Although the river is an important ecological and socio-economic asset, serving as a home to important bird and fish species, and providing water for small-scale irrigation, the health and functionality of the entire system are being threatened by deteriorating water quality. Several sources of pollution including raw sewage run-off from informal settlements, treated wastewater effluent discharges from municipal treatment works, agricultural farmlands, surrounding road and rail networks, and industrial sites were all influencing the water quality of the river and hence the need to assess its health using the SASS5.

4.3. Sampling sites and macroinvertebrates sampling

Four sites within the same ecoregion were selected for the study. Site 1 (33°45′08.4″ S, 25°20′32.6″ E), situated in the upper reaches of the river was the least impacted and thus was chosen as the control site. It has a well diverse range of macroinvertebrate sampling habitats. Site 2

(33°47′29.0″ S, 25°24′26.4″ E) was in the industrial town of Uitenhage, where surrounding impacts include run-off from roads and informal settlements, free-ranging livestock and other agricultural practices. All macroinvertebrate sampling biotopes were adequately represented at the site. Site 2 is situated upstream of the discharge point of the Kelvin Jones wastewater treatment work (WWTW) in the town of Uitenhage. Site 3 (33°47′11.8″ S, 25°25′53.97″ E) is further downstream, but also within the industrial town of Uitenhage, where surrounding impacts include industrial and wastewater effluent discharges, run-off from road and rail networks, and agricultural activities. The Kelvin Jones WWTW is the main pollution source at Site 3. Macroinvertebrate sampling biotopes at Site 3 were also adequate. Site 4 (33°47′34.0″ S, 25°27′58.7″ E) further downstream of Site 3 was situated in the residential town of Despatch. Municipal run-off, sand and gravel mining on the riparian zone were the main impacts at Site 4. Although Site 4 was not as polluted as Site 3, it would have been good to select another site further downstream to monitor for potential system recovery. However, the tidal limit at Perseverance between the estuary and the freshwater section is only a short distance downstream of Site 4. Consequently, it was not possible to select a fifth site further downstream because of likely estuarine effects.

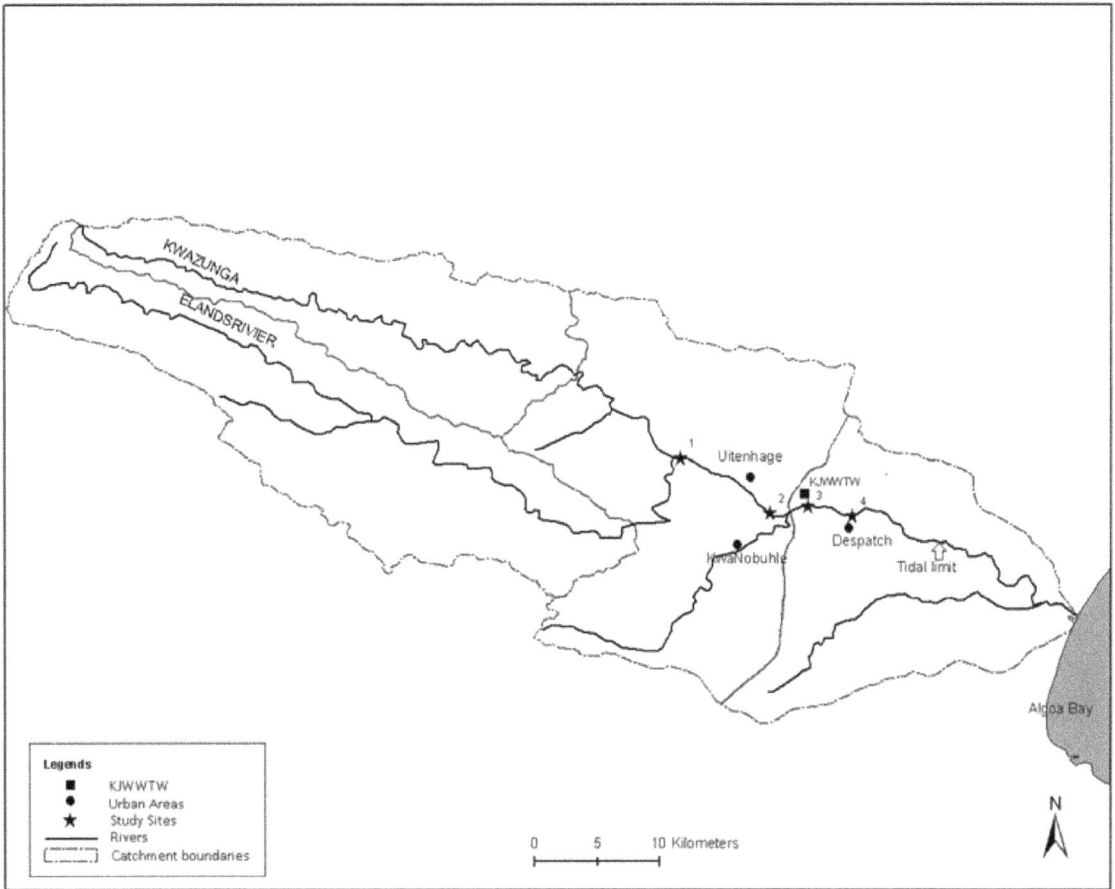

Figure 1. Map of the Swartkops River showing the sampling sites and the relative position of the Kelvin Jones Wastewater Treatment Works.

Macroinvertebrates were sampled using the SASS5 protocol. At each site, over a period of three years, between late August 2009 and September 2012, samples were collected seasonally. A total of eight sampling events were conducted over the sampling period. Macroinvertebrates were collected using a kick net (300 × 300 mm frame, 1000 μm mesh) from three distinct biotope groups: stones (stones-in-and-out-of-current), vegetation (marginal and aquatic vegetation) and sediment (gravel, sand and mud, GSM) as prescribed in the SASS5 protocol.

Sampled macroinvertebrates were tipped into a white rectangular tray, half-filled with river water, and macroinvertebrate families identified on site using identification keys by Gerber and Gabriel [38]. The identified families were recorded on a SASS5 sheet together with their abundance estimates. SASS5 scores, number of taxa and ASPT values were calculated and then interpreted as described in the following section. Time spent on field identification adhered strictly to recommendation in the SASS5 protocol.

4.4. Interpretation of macroinvertebrate data based on the SASS5 protocol for river health

Guidelines developed by Dallas [39] were used for the interpretation of the macroinvertebrate data. The guidelines stipulate range of SASS5 scores and ASPT values indicative of different ecological categories reflective of water quality/river health conditions for the upper and lower areas of each geo-morphological zone in South Africa. The Swartkops River is within the southern eastern coastal belt (lower zone) and the ranges of SASS5 scores and ASPT values for this zone were applied in this study to interpret the SASS5 data in order to determine the Swartkops River health condition (**Table 1**).

4.5. Water sampling and physico-chemical analyses

Basic water physico-chemical analyses were undertaken at each site at the same time when macroinvertebrates were sampled. Dissolved oxygen (DO), electrical conductivity (EC), turbidity, temperature and pH were measured using CyberScan DO 300, CyberScan Con 300, Orbeco-Hellige 966, mercury-in-glass thermometer and CyberScan pH 300 m, respectively. Five-day biochemical oxygen demand (BOD_5) was analysed according to APHA [40].

Ecological category	Water quality category name	Description	Range of SASS5 scores	Range of ASPT values
E/F	Very poor	Seriously/critically modified	<62.9	<5
D	Poor	Largely modified	63–81.9	5.1–5.3
C	Fair	Moderately modified	82–99.9	5.4–5.9
B	Good	Largely natural with few modifications	100–148.9	6.0–7.0
A	Natural	Unmodified	149–180	7.1–8

Table 1. Range of SASS5 scores and ASPT values indicative of the different ecological categories and water quality for the southern eastern coastal belt lower zone ecoregion [39].

4.6. Statistical analysis

One-way analysis of variance (ANOVA) was used to test for differences ($p < 0.05$) in the means of the analysed physico-chemical variables between the four sampling sites. When ANOVA indicated significant differences, a post hoc test, the Tukey's Honestly Significant Different (HSD) test was computed to indicate sites that differed. The basic assumptions of normality and homogeneity of variance were investigated using the Shapiro-Wilk test and the Levene's test, respectively. The nonparametric Kruskal-Wallis multiple comparison test was used to evaluate whether SASS5 scores, ASPT values and the number of taxa differed significantly between the biotope groups. ANOVA and Kruskal-Wallis multiple comparison tests were undertaken using the Statistica software package version 9.

4.7. Results

4.7.1. Water physico-chemical variables

Table 2 shows the mean, standard deviation and range of physico-chemical variables measured during the study period. With the exception of pH and temperature, the measured variables were statistically significantly different between the sampling sites ($p < 0.05$). The lowest value of DO and highest turbidity and BOD_5 values were recorded at Site 3. The Tukey's HSD post hoc test revealed that the mean DO concentration was significantly lower at Site 3 than at Sites 1 and 2. Although pH and temperature were not statistically significantly different between the sampling sites, the highest mean pH and temperature values were at Sites 2 and 3, respectively, and the lowest at Sites 1 and 2, respectively. The Tukey's HSD post hoc test showed that the mean EC concentration was significantly lower at Site 1 than at the rest of the sampling sites and turbidity significantly higher at Site 3. The mean BOD_5 concentrations were significantly higher at Sites 3 and 4 than at Site 1 (**Table 2**).

Variable	Site 1	Site 2	Site 3	Site 4	p value	F value
Dissolved oxygen (mg/l)	6.99 ± 1.15^a (4.73–9.5)	7.4 ± 1.52^a (5.53–9.48)	3.19 ± 1.47^b (1.81–6.36)	4.81 ± 3.01^{ab} (0.9–8.31)	0.001	7.18
pH	6.53 ± 1.11 (4.69–7.75)	7.37 ± 1.11 (5.69–8.99)	7.29 ± 0.42 (6.56–7.9)	7.27 ± 0.56 (6.31–8.01)	0.201	1.65
Temperature (°C)	17.48 ± 5.46 (7.31–24.0)	17.27 ± 7.17 (6.11–27.3)	20.88 ± 3.29 (14.3–25.2)	18.9 ± 4.14 (12.2–24.0)	0.415	0.98
Electrical conductivity (mS/m)	32.45 ± 17.74^a (8.23–62.0)	160.75 ± 146^b (30–460)	262.51 ± 76.14^b (154.8–333)	259.63 ± 56.28^b (171–354)	0.000	22.57
Turbidity (NTU)	5.3 ± 2.22^a (3.0–10.1)	6.33 ± 2.44^a (3.0–11.2)	72.7 ± 102.36^b (10.5–320)	7.08 ± 8.06^a (2.2–26)	0.000	15.67
BOD_5 (mg/l)	4.62 ± 1.45^a (2.16–6.86)	8.25 ± 4.33^{ab} (4.58–16.68)	14.54 ± 3.57^c (8.32–20.62)	11.77 ± 5.28^{bc} (2.24–22.94)	0.002	13.50

Table 2. Mean ± standard deviation and range (in parenthesis) of the physico-chemical variables ($n = 8$) in the Swartkops River during the study period (August 2009–September 2012). p and F values are indicated by ANOVA. Different superscript letters per variable across sites indicate significant differences ($p < 0.05$) revealed by Tukey's HSD post hoc test. The same superscript letter between sites per variable indicates no significant differences ($p > 0.05$).

4.7.2. Assessing the Swartkops River health using the South African Scoring System version 5 (SASS5)

The interpretation of the SASS5 results were based on the range of SASS5 scores and ASPT values reflecting ecological categories A, B, C, D and E/F indicative of natural, good, fair, poor and very poor water quality conditions, respectively (**Table 1**). The SASS5 scores and ASPT values revealed that the Swartkops river health conditions differed between the sampling sites. Seasonally, with the exception of the autumn and spring (2012) collections, SASS5 scores at Site 1 indicated the B ecological category indicative of good water quality condition (**Figure 2**). The ASPT values on the other hand, in all the sampling seasons, indicated the C ecological category for Site 1, suggesting that the water quality at Site 1 was fair (**Figure 3**). The numbers of taxa vary slightly between the sampling seasons at Site 1 with more taxa occurring in spring (2012) (**Figure 4**). Overall, the SASS5 score showed good water quality (ecological category B) for Site 1, but the ASPT value indicated that the water quality condition at the site was fair (ecological category C) (**Figure 5**).

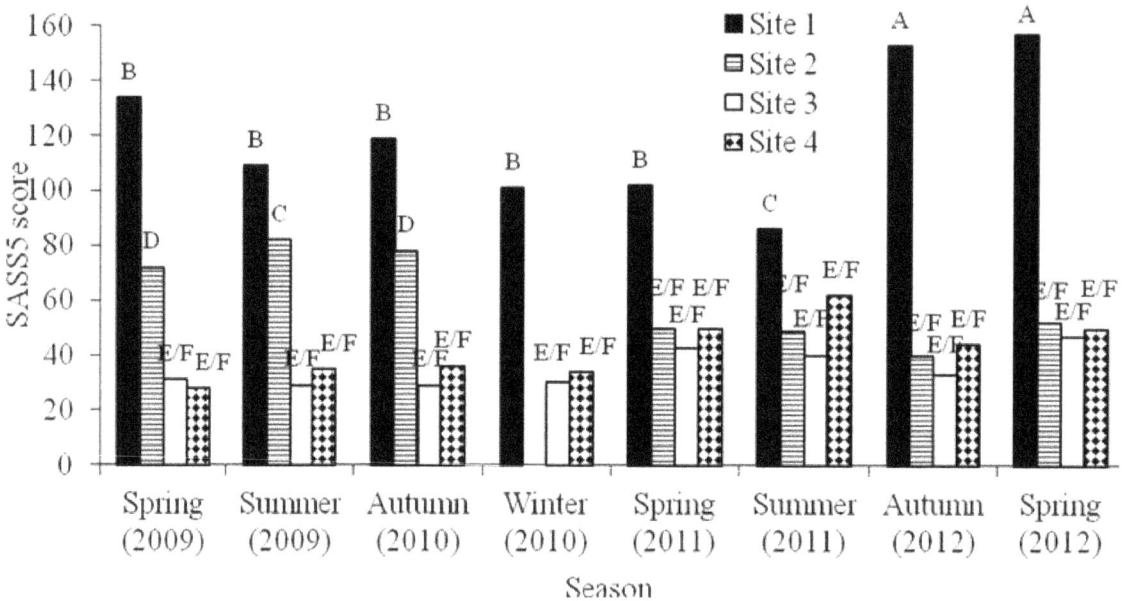

Figure 2. Seasonal variations for the South African Scoring System version 5 (SASS5) score at the four sampling sites in the Swartkops River during the study period (August 2009–September 2012). The ecological categories: A (natural water quality), B (good water quality), C (fair water quality), D (poor water quality) and E/F (very poor water quality) are indicated on the bars.

At Site 2, SASS5 scores indicated the D ecological category, that is, poor water quality in spring (2009) and in autumn (2010), while in summer (2009), it revealed the C category indicative of fair water quality (**Figure 2**). During the rest of the sampling events, SASS5 scores revealed the E/F ecological category indicating very poor water quality. Although the SASS5 scores reflected other ecological categories in addition to the E/F for Site 2, the ASPT values consistently showed that Site 2 was in the E/F ecological category (**Figure 3**). Although the number of taxa did not vary significantly between the sampling seasons at Site 2, the highest number of taxa (20) was recorded during autumn (2010). At Sites 3 and 4, SASS5 scores and ASPT values revealed the E/F ecological category (very poor water quality) throughout the sampling seasons. The overall lowest number of taxa (8) in the river was recorded at Site 3 in winter 2010 (**Figure 4**).

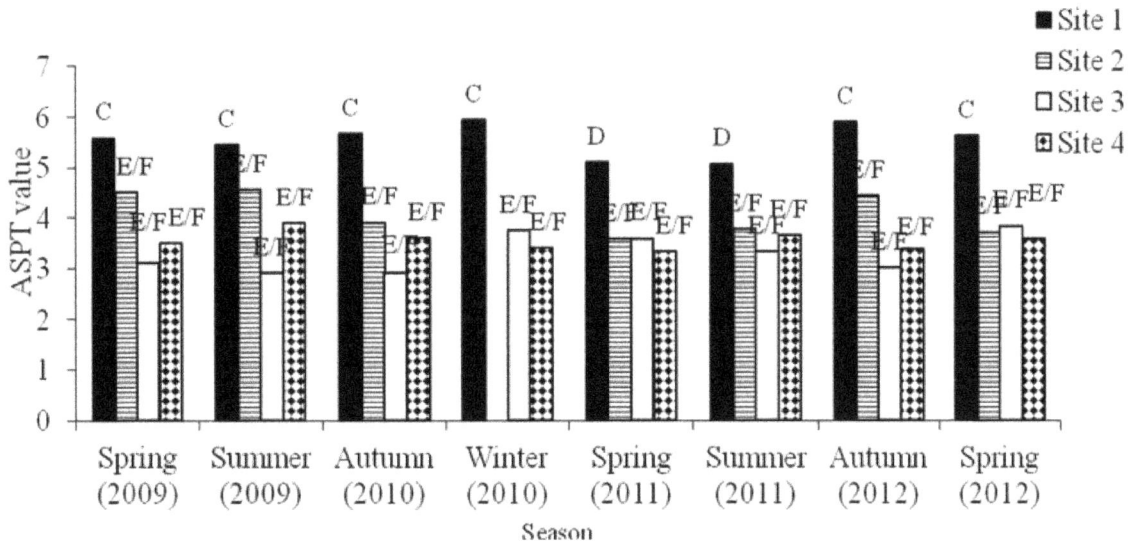

Figure 3. Seasonal variations for the average score per recorded taxon (ASPT) at the four sampling sites in the Swartkops River during the study period (August 2009–September 2012). The ecological categories: C (fair water quality), D (poor water quality) and E/F (very poor water quality) are indicated on the bars.

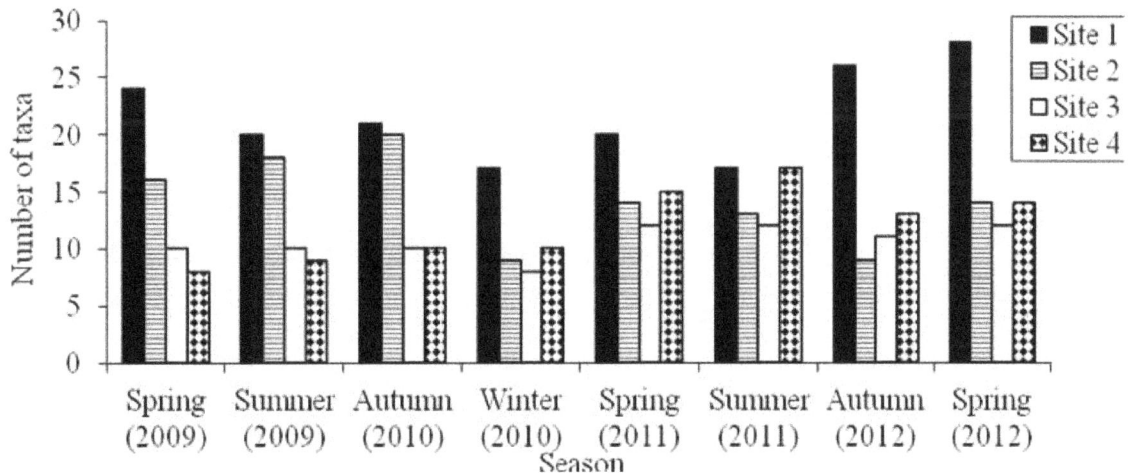

Figure 4. Seasonal variations for the number of taxa at the four sampling sites in the Swartkops River during the study period (August 2009–September 2012).

4.7.3. Comparing SASS5 scores, ASPT values and the numbers of taxa between the sampling biotopes (stone, vegetation and GSM)

The vegetation and stone biotope had higher SASS5 scores, ASPT values and numbers of taxa than the GSM biotope at Site 1 (**Figure 6**). The Kruskal-Wallis multiple comparison test revealed that SASS5 scores were significantly higher for the vegetation than for the GSM biotope at Site 1 ($p < 0.05$; KW-H = 7.21). Similarly, at Site 2, SASS5 scores were significantly higher for the vegetation than for the GSM biotope ($p < 0.05$; KW-H = 10.13), and though the stone had higher SASS5 scores, they were not statistically higher than the scores recorded for the GSM biotope. The pattern described for Site 2 was similar to those observed for Sites 3 and 4 where the SASS5 scores were significantly higher for the vegetation biotope than the stone and GSM biotopes (Site 3: $p < 0.05$; KW-H = 40.44), (Site 4: $p < 0.05$; KW-H = 18.14).

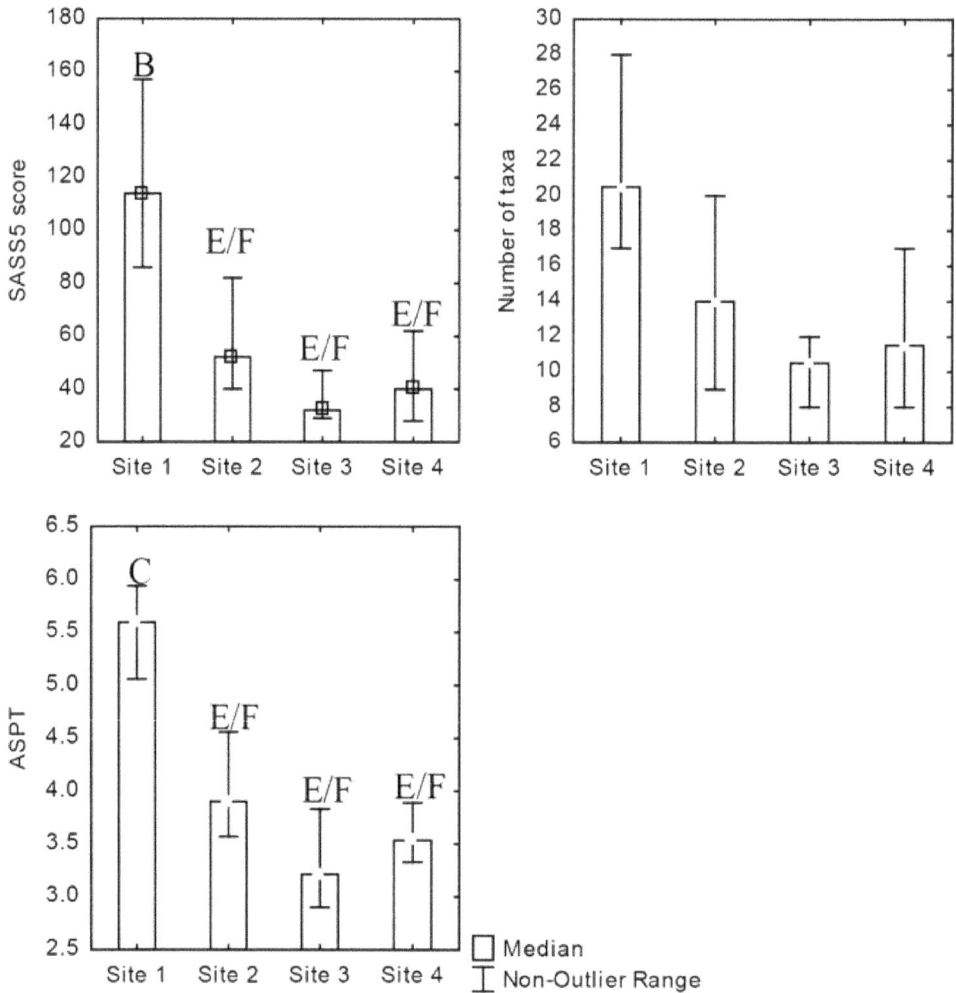

Figure 5. Summary of the SASS5 scores, number of taxa and ASPT values at the four sampling sites in the Swartkops River during the study period (August 2009–September 2012). The overall ecological categories: B (good water quality), C (fair water quality), D (poor water quality) and E/F (very poor water quality) are indicated on the bars.

The average score per recorded taxon (ASPT) values were similar between the three biotopes at Site 1, but at Site 2, the ASPT values were significantly higher for the vegetation than the GSM biotope ($p < 0.05$; KW-H = 9.45). The vegetation had significantly higher ASPT values than the stone and GSM biotopes at Sites 3 ($p < 0.05$; KW-H = 26.9) and 4 ($p < 0.05$; KW-H = 14.25).

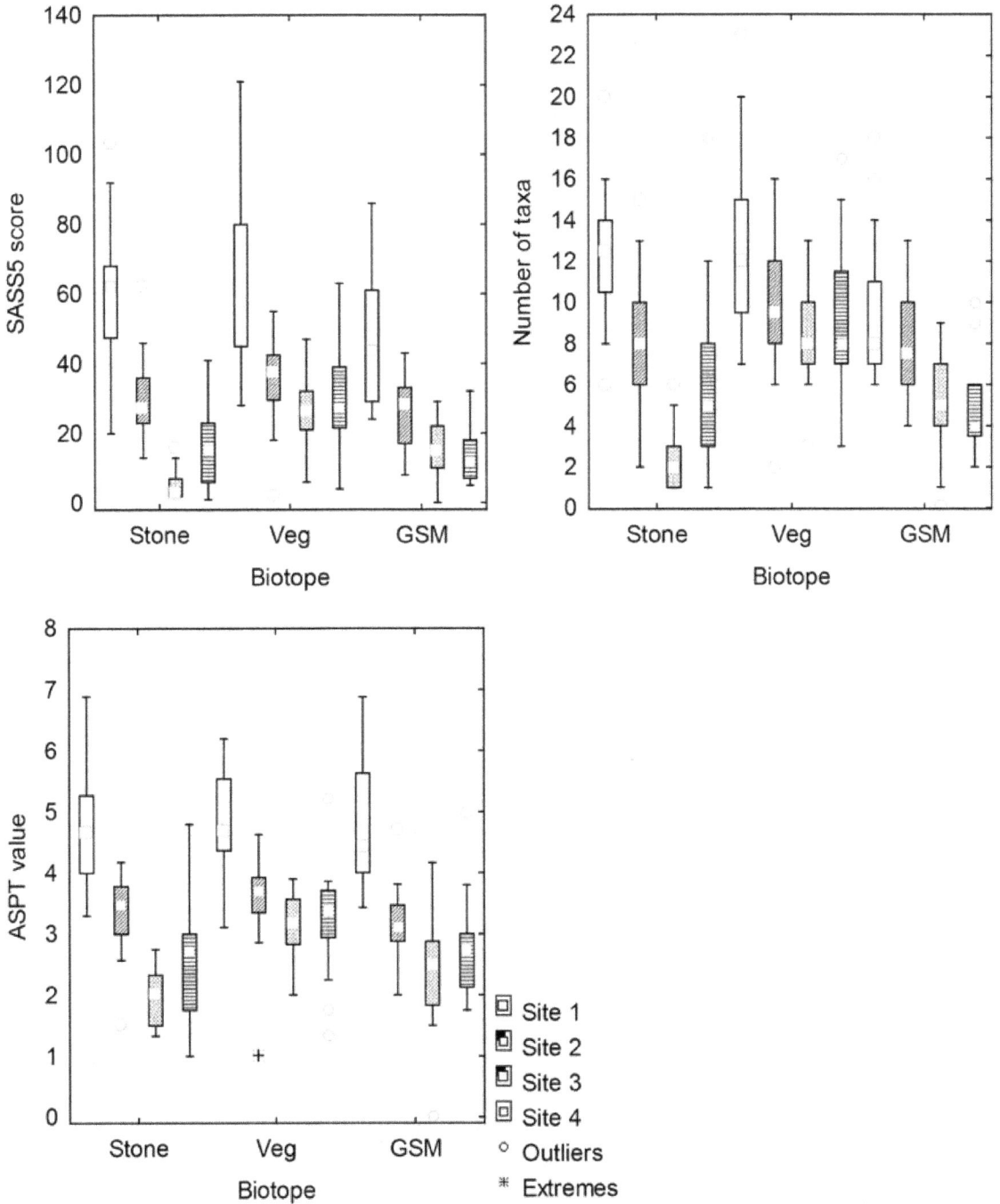

Figure 6. Median (small square), inter-quartile ranges (box), non-outlier ranges (bars) for SASS5 scores, numbers of taxa and ASPT values recorded per biotope at the four sampling sites in the Swartkops River during the study period (August 2009–September 2012).

Stone and vegetation biotopes supported significantly higher numbers of taxa than the GSM biotope at Site 1 ($p < 0.05$; KW-H = 11.89), but at Site 2, only the vegetation supported significantly higher numbers of taxa than the GSM ($p < 0.05$; KW-H = 7.23). More taxa were recorded on the vegetation and GSM biotopes than on the stone biotopes at Site 3. The Kruskal-Wallis multiple comparison test indicated that the numbers of taxa for the stone biotope were significantly lower than the taxa recorded for the vegetation and GSM ($p < 0.05$; KW-H = 40.44) at Site 3. At Site 4, the stone and vegetation supported more taxa, but only the numbers of taxa supported by the vegetation biotope were significantly higher than the values recorded for the GSM ($p < 0.05$; KW-H = 16.27).

4.8. Discussion

The ecosystem approach takes into account biodiversity conservation and therefore prioritises the protection of biodiversity as well as the sustainable use of water resources and the associated ecosystems. In the case study provided, the South African Scoring System version 5 (SASS5) was used to evaluate the health of the Swartkops River. In South Africa, SASS5 is one of the tools that contribute ecological information for the determining the ecological reserve and setting Resource Quality Objectives (RQOs). The SASS5 results indicated that water quality in the Swartkops River was critically modified at Sites 3 and 4 throughout the sampling period and the numbers of taxa occurring at these sites were significantly reduced compared to those occurring at Sites 1 and 2. Sites 3 and 4 were situated downstream of a WWTWs, which influenced the health of the river. The values of the measured physico-chemical variables at these sites, that is, Sites 3 and 4, provided evidence for negative impact arising from the discharges of wastewater effluents. For example, at Sites 3 and 4, higher values of turbidity and EC concentrations and lower DO concentrations were recorded. Since highly sensitive taxa have higher scores in the SASS5 sheet, oxygen depletion could easily affect the occurrence and distribution of these taxa. Therefore, it was expected that sites with low concentration of DO would experience the disappearance of sensitive taxa and the dominance of tolerant taxa, and hence, the critically modified health conditions recorded at Sites 3 and 4. In addition to lower DO concentrations at the downstream sites, the elevated turbidity level recorded at Site 3 could be detrimental to oxygen-sensitive biota as decomposition of solids with high organic content could lead to oxygen depletion, as was evident at Sites 3 and 4. The majority of the highly sensitive taxa on the SASS5 sheet use external gills for respiration. Highly turbid water is likely to impact on the breathing apparatus of external gill-bearing organisms, which can then lead to clogging [41]. The river health condition at Site 2, which was upstream of the effluent discharge point, but still situated within the urban and industrial town of Uitenhage, was mostly in the range of fair and very poor conditions. Diffuse pollution sources on the river catchments were the main contributors to deteriorating river health recorded at this site. Site 1, which was used as the control site, had conditions mostly in the good and fair categories. The implication was that the control site had some sensitive taxa, which had disappeared from the impacted sites.

The number of taxa, SASS5 scores and ASPT values were highest mostly in the stone and vegetation biotopes and differed significantly between the three biotopes. These differences could be attributed to differences in hydraulic, substrate and thermal conditions between the three biotopes. The stone and vegetation biotopes are morphologically complex and more stable than the GSM biotope and are therefore more likely to support more food and space resources, and thus more macroinvertebrate families leading to increased SASS5 scores and ASPT values. These results are in agreement with those of Dallas [42] who reported that the stone and vegetation biotopes supported more macroinvertebrate families and higher SASS5 score and ASPT values than the GSM biotope. It is therefore important to sample all available biotopes to capture a wider range of biodiversity when undertaken aquatic biomonitoring.

In summary, the deteriorating environmental water quality in the Swartkops River has impacted on the macroinvertebrate assemblages particularly at the downstream sites. This was expected because of the ranges of impacts these sites receive which include industrial and sewage effluent discharges, run-off from informal settlement and agricultural activities such as livestock farming. Water quality at Site 1 which was used as the control site in this study was indicated as good and fair by the SASS5 score and ASPT value, respectively. This is a cause for concern as the results showed that macroinvertebrates at this site were experiencing noticeable impacts. Overall, both the physico-chemical variable analysis and the biotic index results revealed that the Swartkops River was deteriorating in quality as it flowed downstream, indicating the need for an urgent management intervention.

5. Conclusion

In this chapter, the ecosystem-based approach to managing water quality was critically reviewed with a clear focus on environmental water quality (EWQ). The three pillars to EWQ were discussed and their contributions and limitation highlighted. Of particular interest is that, in this chapter, the relevance of the EWQ approach was discussed with respect to its application to water resources management in South Africa. It is argued that the EWQ is an integrative approach for sound and sustainable management of water quality. The biomonitoring case study illustrated the utility of one of the three pillars of the EWQ approach.

Acknowledgements

The South African National Research Foundation (NRF) is acknowledged for providing postdoctoral grant (Grant no.: 88517) to Dr. O.N. Odume for this research. The Carnegie Corporation of New York through the Regional Initiative in Science and Education (RISE) is also acknowledged for doctoral bursary for this work. Rhodes University Research Committee (RC) is acknowledged for a research grant.

Author details

Oghenekaro Nelson Odume

Address all correspondence to: nelskaro@yahoo.com

Unilever Centre for Environmental Water Quality, Institute for Water Research, Rhodes University, Grahamstown, South Africa

References

[1] Bunn, S.E. (2016) Grand challenge for the future of freshwater ecosystems. *Frontiers in Environmental Science* **4(21)**: 1–4.

[2] Woznicki, S.A., Nedadhashemi, A.P., Tang, Y., and Wang, L. (2016) Large-scale climate change vulnerability assessment of stream health. *Ecological Indicators* **69**: 578–598.

[3] Masese, F.O., Omokoto, J.O., and Nyakeya, K. (2013) Biomonitoring as a prerequisite for sustainable water resources: a review of current status, opportunities and challenges to scaling up in East Africa. *Ecohydrology and Hydrobiology* **13(3)**: 173–191.

[4] Brown, P.G., and Schmidt, J.J. (2010) An ethic of compassionate retreat. In: Brown, P.G. and Schmidt, J.J. editors *Water ethics—foundational readings for students and professionals*. Island Press, Washington. p. 265–286.

[5] Department of Water Affairs (2013) National water resource strategy. Second edition. Department of Water Affairs, Pretoria, South Africa.

[6] Pollard, S. and du Toit, D. (2008) Integrated water resource management in complex systems: how the catchment management strategies seek to achieve sustainability and equity in water resources in South Africa. *Water SA* **34(6)**: 671–679.

[7] Doulgeris, C., Georgiou, P., Papadimos, D., and Papamichail, D. (2012) Ecosystem approach to water resources management using the Mike 11 modeling system in the Strymonas River and Lake Kerkini. *Journal of Environmental Management* **94**: 132–143.

[8] Thoms, M.C., and Sheldon, F. (2002) An ecosystem approach for determining environmental water allocations in Australian dryland river systems: the roe of geomorphology. *Geomorphology* **47**: 153–168.

[9] King, J., and Pienaar, H. (2011) Sustainable use of South Africa's inland waters. WRC Report No. TT 491/11. Water Research Commission, Pretoria, South Africa.

[10] DWAF—(Department of Water Affairs and Forestry) (2008b) Draft regulations for the establishment of a water resource classification system. Government Gazette No. 31417, Pretoria, South Africa.

[11] Odume, O.N. (2014) Macroinvertebrates-based biomonitoring and ecotoxicological assessment of deteriorating environmental water quality in the Swartkops River, South Africa. PhD thesis, Rhodes University, Grahamstown, South Africa.

[12] Scherman, P.A., Muller, W.J., and Palmer, C.G. (2003) Links between ecotoxicology, biomonitoring and water chemistry in the integration of water quality into environmental flow assessments. *River Research and Applications* **19**: 1–11.

[13] Palmer, C.G., Berold R., and Muller W.J. (2004) Environmental water quality in water resource management. WRC Report No.TT 217/04, Water Research Commission, Pretoria, South Africa.

[14] Hohls, B.C., Silberbauer, M.J., Kühn, A.L., Kempster, P.L., and van Ginkel, C.E. (2002) National water resource quality status report: inorganic chemical water quality of surface water resources in SA—the big picture. Report No. N/0000/REQ0801. ISBN No. 0-621-32935-5. Institute for Water Quality Studies, Department of Water Affairs and Forestry, Pretoria, South Africa. http://www.dwaf.gov.za/iwqs/water_quality/NCMP/ReportNationalAssmt3c.pdf [Accessed 28 May, 2012]

[15] Extence, C.A., Chadd, R.P., England, J., Dunbar, M.J., Wood, P.J., and Taylor, E.D. (2013) The assessment of fine sediment accumulation in rivers using macroinvertebrate community response. *River Research and Applications* **29**: 17–55.

[16] Bonada, N., Prat, N., Resh, V.H., and Statzner, B. (2006) Development in aquatic insect biomonitoring: a comparative analysis of recent approaches. *Annual Review of Entomology* **51**: 495–523.

[17] Department of Water Affairs and Forestry (2003) National Aquatic ecosystem biomonitoring programme—compiling State of the rivers report and posters: a manual. NAEBP Report Series No. 17. DWAF, Pretoria, South Africa.

[18] Kleynhans, C.J., and Louw, M.D. (2008) River ecoclassification: manual for ecostatus determination (version 2)—Module A: EcoClassification and EcoStatus determination. Joint Water Research Commission and Department of Water Affairs and Forestry Report. WRC Report No TT 329/08. Water Research Commission, Pretoria, South Africa.

[19] Kleynhans, C.J. (2008) River ecoclassification: manual for ecostatus determination (version 2)—Module D volume 1: Fish Response Assessment Index. Joint Water Research Commission and Department of Water Affairs and Forestry Report. WRC Report No TT 330/08. Water Research Commission, Pretoria, South Africa.

[20] Thirion, C. (2008) River ecoclassification: manual for ecostatus determination (version 2)—Module E: macroinvertebrate response assessment index in. Joint Water Research Commission and Department of Water Affairs and Forestry Report. Water Research Commission, Pretoria, WRC Report No. TT 332/08.

[21] Pace, G., Bella, V.D., Barile, M., Andreani, P., Mancini, L., and Belfiore, C. (2012) A comparison of macroinvertebrate and diatom responses to anthropogenic stress in a small sized volcanic siliceous streams of central Italy (Mediterranean Ecoregion). *Ecological Indicators* **23**: 544–554.

[22] Murphy, J.F., Davy-Bowker, J., McFarland, B., and Ormerod, S.J. (2013) A diagnostic biotic index for assessing acidity in sensitive streams in Britain. *Ecological Indicators* **24**: 562–572.

[23] Rosenberg, D.M., and Resh, V.H. editors. (1993) *Freshwater biomonitoring and benthic macroinvertebrates*, p. 488. Chapman and Hall One Penn Plaza, New York, NY.

[24] Dickens, C.W.S., and Graham, P.M. (2002) The South African Scoring System (SASS) version 5 rapid bioassessment method for rivers. *African Journal of Aquatic Science* **27**: 1–10.

[25] Walley, W.J., and Hawkes, H.A. (1996) A computer-based reappraisal of the biological monitoring working party scores using data from the 1990 river quality survey of England and Wales. *Water Research* **30**: 2086–2094.

[26] Baptista, D.F., Bus, D.F., Egler, M., Giovanelli, A., Silveira, M.P., and Nessimian, J.L. (2007) A multimetric index based on benthic macroinvertebrates for evaluation of Atlantic Forest streams at Rio de Janeiro State, Brazil. *Hydrobiologia* **575**: 83–94.

[27] Jorgensen S.E., Costanza, R., and Xu, F.L. (2005) Handbook of ecological indicators for assessment of ecosystem health. CRC Press Boca Raton, USA.

[28] Rand, G.M., (Ed.) (1995) Fundamentals of aquatic toxicology—effects, environmental fates, and risk assessment (2nd edn), p. 1148. Taylor and Francis, Washington D.C., USA.

[29] Mensah, P.K., Muller, W.J., and Palmer, C.G. (2011) Acute toxicity of Roundup® herbicide to three life stages of the freshwater shrimp *Caridina nilotica* (Decapoda: Atyidae). *Physics and Chemistry of the Earth* **36**: 905–909.

[30] Moiseenko, T.I. (2008) Aquatic ecotoxicology: theoretical principles and practical application. *Water Resources* **35(5)**: 530–541.

[31] Schmitt-Jansen, M., Veit, U., Dudel, G., and Altenburger, R. (2008) An ecological perspective in aquatic ecotoxicology: approaches and challenges. *Basic and Applied Ecology* **9**: 337–345.

[32] Ledger, M.E., Harris, R.M.L., Armitage, P.D., and Milner, A.M. (2009) Realism of model ecosystems: an evaluation of physicochemistry and macroinvertebrate assemblages in artificial streams. *Hydrobiologia* **617**: 91–99.

[33] Odum, E.P. (1984) The mesocosm. *Bioscience* **34**: 558–562.

[34] Buikema, A.L., and Voshell, J.R. (1993) Toxicity studies using freshwater benthic macroinvertebrates. In: Rosenberg, D.M., and Resh, V.H. (Eds.), *Freshwater biomonitoring and benthic macroinvertebrates*. Chapman and Hall one penn plaza, New York, NY. pp. 344–398.

[35] Hill, I.R., Heimbach, F., Leeuwangh, P., and Matthiessen, P. (Eds.) (1994) Freshwater field tests for hazard assessment of chemicals, p. 561. CRC Press, Boca Raton, FL.

[36] Belanger, S.E. (1997) Literature review and analysis of biological complexity in model stream ecosystems: influence of size and experimental design. *Ecotoxicology and Environmental Safety* **36**: 1–16.

[37] Choung, C.B., Hyne, R.V., Stevens, M.M., and Grant C.H. (2013) The ecological effects of a herbicide-insecticide mixture on an experimental freshwater ecosystem. *Environmental Pollution* **172**: 264–274.

[38] Gerber, A., and Gabriel, M.J.M. (2002) Aquatic invertebrates of South African Rivers—field guide. Resource quality services, Department of Water Affairs, Pretoria.

[39] Dallas, H.F. (2007) River health programme: South African scoring system (SASS) data interpretation guidelines. Institute of Natural Resources and Department of Water Affairs and Forestry, Pretoria, South Africa. Available online: http://safrass.com/reports/SASS% 20Interpretation%20Guidelines.pdf Accessed: 18 October, 2010.

[40] APHA—(American Public Health Association) (1992) Standard methods for the examination of water and waste water (18th edn). APHA, Washington DC, USA.

[41] Bilotta, G.S., and Brazier, R.E. (2008) Understanding the influence of suspended solids on water quality and aquatic biota. *Water Research* **42**: 2849–2861.

[42] Dallas, H.F. (2007) The influence of biotope availability on macroinvertebrate assemblages in South African rivers: implications for aquatic bioassessment. *Freshwater Biology* **52**:370–380.

PERMISSIONS

All chapters in this book were first published in WQ, by InTech Open; hereby published with permission under the Creative Commons Attribution License or equivalent. Every chapter published in this book has been scrutinized by our experts. Their significance has been extensively debated. The topics covered herein carry significant findings which will fuel the growth of the discipline. They may even be implemented as practical applications or may be referred to as a beginning point for another development.

The contributors of this book come from diverse backgrounds, making this book a truly international effort. This book will bring forth new frontiers with its revolutionizing research information and detailed analysis of the nascent developments around the world.

We would like to thank all the contributing authors for lending their expertise to make the book truly unique. They have played a crucial role in the development of this book. Without their invaluable contributions this book wouldn't have been possible. They have made vital efforts to compile up to date information on the varied aspects of this subject to make this book a valuable addition to the collection of many professionals and students.

This book was conceptualized with the vision of imparting up-to-date information and advanced data in this field. To ensure the same, a matchless editorial board was set up. Every individual on the board went through rigorous rounds of assessment to prove their worth. After which they invested a large part of their time researching and compiling the most relevant data for our readers.

The editorial board has been involved in producing this book since its inception. They have spent rigorous hours researching and exploring the diverse topics which have resulted in the successful publishing of this book. They have passed on their knowledge of decades through this book. To expedite this challenging task, the publisher supported the team at every step. A small team of assistant editors was also appointed to further simplify the editing procedure and attain best results for the readers.

Apart from the editorial board, the designing team has also invested a significant amount of their time in understanding the subject and creating the most relevant covers. They scrutinized every image to scout for the most suitable representation of the subject and create an appropriate cover for the book.

The publishing team has been an ardent support to the editorial, designing and production team. Their endless efforts to recruit the best for this project, has resulted in the accomplishment of this book. They are a veteran in the field of academics and their pool of knowledge is as vast as their experience in printing. Their expertise and guidance has proved useful at every step. Their uncompromising quality standards have made this book an exceptional effort. Their encouragement from time to time has been an inspiration for everyone.

The publisher and the editorial board hope that this book will prove to be a valuable piece of knowledge for researchers, students, practitioners and scholars across the globe.

LIST OF CONTRIBUTORS

Benias C. Nyamunda
Midlands State University, Manicaland College of Applied Sciences, Department of Chemical and Processing Engineering, Mutare, Zimbabwe

John Okapes Joseph, Isaac W. Mwangi, Sauda Swaleh, Ruth N. Wanjau and Manohar Ram
Department of Chemistry, Kenyatta University, Nairobi, Kenya

Jane Catherine Ngila
Department of Chemical Technology, University of Johannesburg, South Africa

Salmiati and Mohd Razman Salim
Department of Environmental Engineering, Faculty of Civil Engineering, University of Technology, Johor Bahru, Johor, Malaysia
Center for Environmental Sustainability and Water Security (IPASA), RISE, University of Technology, Johor Bahru, Johor, Malaysia

Nor Zaiha Arman
Department of Environmental Engineering, Faculty of Civil Engineering, University of Technology, Johor Bahru, Johor, Malaysia

Bloodless Dzwairo
Durban University of Technology, Department of Civil Engineering (Midlands), Durban, South Africa

Munyaradzi Mujuru
University of Limpopo, Department of Water and Sanitation, Faculty of Science and Agriculture, Sovenga, Medunsa, Polokwane, South Africa

Lydia Bondareva and Valerii Rakitskii
Federal Scientific Center of Hygiene named after F.F. Erismana, Moscow, Russia

Ivan Tananaev
Far Eastern Federal University, Vlodivostok, Russia

Ebru Karnez
Faculty of Agriculture, Cukurova University, Adana, Turkey

Hande Sagir and Hayriye Ibrikci
Soil Science and Plant Nutrition Department, Cukurova University, Adana, Turkey

Matjaž Gavan and Marina Pintar
Department for Agronomy, Biotechnical Faculty, University of Ljubljana, Ljubljana, Slovenia

Muhammed Said Golpinar and Mahmut Cetin
Department of Irrigation Engineering, Faculty of Agriculture, Cukurova University, Adana, Turkey

Mehmet Ali Akgul
The Sixth Regional Directorate of State Hydraulic Works (DSI), Adana, Turkey

Stefania Gheorghe, Catalina Stoica, Gabriela Geanina Vasile, Mihai Nita-Lazar, Elena Stanescu and Irina Eugenia Lucaciu
National Research and Development Institute for Industrial Ecology – ECOIND, Bucharest, Romania

Oghenekaro Nelson Odume
Unilever Centre for Environmental Water Quality, Institute for Water Research, Rhodes University, Grahamstown, South Africa

Index